Welcome to Florida

Welcome To Florida

A Different Look At The Sunshine State

Robert Allen Morris

This book is available at www.amazon.com

Orchid Springs Publishing, LLC
329 N. Park Ave., Floor 2
Winter Park, FL 32789

ISBN: 978-0-9963190-7-2 (paperback)
ISBN: 978-0-9963190-8-9 (eBook)

Contents

Acknowledgements

Anna Maria and Valencia Morning paintings, courtesy Paul Schulz, *www.paulschulz.com*. Turpentine Face, courtesy **Pictorial Album of the Navel Stores Industry, Library and Archives**, Forest History Society, Durham, NC, USA. Photo of Florida Cowboy by Carlton Ward, Jr. *www.carltonward.com*.

Introduction

I'm a Florida native, born in Tampa on October 3, 1952, something that's rare in Florida these days, I'm also an agricultural economist, educated at the University of Florida, with 44 years of experience in Florida agribusiness. My biographical sketch is at the end of this book. Other than two years when I worked for The Coca-Cola Company and lived in Atlanta, I have spent my entire life in Florida. Writing this book has reminded me how fortunate I am to have done that.

I've watched the sunrise far into the woods on horseback as the ground mist lifts like wisps of smoke and creates a heavenly green glow in the pine forest and lights up the sedge with a tawny gold, the fragrance of alpine ever-present, the woodland filled with the musical calls of mockingbirds, the raspy noise of blue jays, and crows cawing in the distance. I've seen a newborn calf rise for the first time on wobbly legs, listened to the melodic sound of hounds chasing a wild hog or bobcat in the woods, and smelled bacon cooking over a campfire mingled with the scent of freshly ground coffee and wood smoke. I've paddled a canoe down Florida's wilderness rivers and watched egrets, blue herons and white ibis feeding in the river; alligators silently gliding through the murky shallows stalking and catching lethargic river turtles as they lay sunning on a branch or root near the water. At dawn and dusk I've seen white tailed deer come to drink at the water's edge, often chasing away the occasional raccoon washing its dinner of river mussels or shiners, small silver-colored fish, in the water. And flocks of wild turkeys roosted in cypress swamps bordering the rivers, flying up to their roost just before dark each night and back down to feed at dawn. I've smelled the perfume of citrus trees in full bloom on a crisp March morning and tasted sweet Valencia orange juice straight out of the juice finishers before its pasteurized. I've experienced the exhilarating thrill of churning out into the open ocean on a fishing boat, enjoying the aroma of the sea and a beautiful morning horizon as a day of fishing awaits; and the gorgeous Florida sunset when the red sun sinks its way slowly into the ocean as day turns into night. Someone once said, "If I had a million bucks, I'd move to Florida." Living in Florida has been worth much more than a million bucks to me.

Today, three of Florida's leading industries, in order of economic importance, are agriculture, tourism and construction. With a 2022 population of 22.2 million, Florida is now the third most populous state in the United States, behind California and Texas. Moreover, an estimated 138 million tourists come to Florida each year, and almost 900 people move to Florida each day, many to retire. But this wasn't always so.

My father was killed in an accident on a tractor when I was seven years old, and my mother never remarried. I was raised in the small town of MacClenny, which is approximately 30 miles west of Jacksonville and about 6 miles south of the Florida/Georgia state line. MacClenny is also the county seat of Baker County. When I was born, there were only 6,313 people living in Baker County and Florida's population was only 2.8 million. At that time, eight of Florida's leading agricultural industries were turpentine, timber, cattle, commercial fishing, citrus, vegetables, sugarcane, and wholesale landscape nurseries. Let's look at the way these industries were, and the way they are now.

The Turpentine Industry

The turpentine industry, also called the naval stores industry, derives its name from the use of tar and pitch for waterproofing the hulls of early wooden sailing vessels. Its principal products were tar, pitch, rosin and turpentine, produced from gum extracted from pine trees. By the turn of the twentieth century, with the advent of steel hulls on ships and decreased need for tar and pitch, the gum was distilled into two products: rosin and turpentine. Turpentine was used for fuel in lanterns and stoves, and both products were used in the making of soaps, axle grease, paint thinners, solvents, roofing materials, linoleum, explosives, plasters, disinfectants, insecticides, and other products.

My grandpa Morris and my grandpa Cook spent their entire careers as first turpentine woods riders and then general managers of turpentine camps. My mother was raised in turpentine camps and so was my father, who was a woods rider before they were married. The turpentine industry was one of the oldest in the U.S., dating back to Jamestown in 1608. It was in its zenith in the abundant longleaf and slash pine forests of the Southeastern United States from the 1830's until the 1930's. By the1970's, increasing labor costs, development of new technologies, and availability of synthetic substitutes for turpentine led to the end of the navel stores industry.

The following draws not only from experiences told to me by my grandfathers, but also on *Tapping the Pines* by Robert Outland, and part of the summer of 1967 when I worked for the Southern Resin Chemical Company.

Courtesy USDA Forest Service

A strip of bark ¾ inch wide (a streak) was cut from part of the trunk of a pine tree, called the face, with a tool called a hack. This was called chipping. The tree produces and secrets gum to heal its wound, and tin aprons were nailed to the tree to direct the flow of gum into cups or troughs. About every 7-10 days, another streak was cut into the face of the tree just above the last streak. This chipping process was continued over the 4-6 years that the tree would produce gum until the face grew above head height, and a long-handled hack was used to extend the face on up about 8-10 feet high. When cups filled with the gum, it was dipped from the cups into a wooden barrel hauled on a wagon. It was then hauled to the distillery where it was distilled into rosin and turpentine.

Turpentine operations were organized around the turpentine camp. Due to the isolated locations of the large forest tracts, most camps were located many miles from the nearest paved road and could be reached only by a dirt or rutted company road winding through murky swamps and dense pine forests to the remote camp. Thus, turpentine camps were self-sufficient communities, usually with about 40 to 60 people. They contained dwellings that housed the workers and their families, a house and associated buildings for the general manager and his family, a commissary or general store, a distillery, and a combination church-school. Some also contained a juke joint, typically operated on Saturday nights by an enterprising worker, where workers could drink moonshine, dance and gamble. The general manager of the turpentine camp was responsible for the entire turpentine operation. He directed field operations through his woods riders, who supervised chipping trees and dipping gum from horseback, and was enforcer of the law and order, director of the commissary and distillery, pay master, and make-shift doctor.

One man could attend a "crop" of 10,000 trees spread over about 100 acres. Woods riders were an integral part of turpentine operations.

My grandfather, William Fount Cook, in 1910 at the age of 24. The enterprising young man was already a turpentine camp manager and ready to find a wife. He did – my grandmother, Abigale Camelia Stephens, from Daviston, Alabama. They moved with the turpentine business, from Alabama where my mother was born in 1914, to Georgia where her oldest brother Willard was born in 1916, to Florida where her youngest brother Wildon was born in 1930.

They inspected from horse back the placing of the tin aprons and cups on the trees, chipping, dipping the gum into barrels, and later in the year, collecting scrape. They kept track of how many trees each worker completed daily. And they ensured not only that all tasks were completely and carefully executed, but also that tools and equipment were maintained in good condition and that workers got to and from the correct production sites. Tally men had important jobs as well because they recoded the number of trees each worker completed when they completed them. When a worker finished his task at one face, he called out his individual code loudly enough for the tallyman to hear him up to several hundred yards away. The different codes shouted throughout the forest reportedly took on a rhythmic chanting sound. Each tallyman could record the work of about 10-12 hands. The tallyman needed to be competent, trustworthy, and of good hearing, but age and strength mattered little.

Courtesy the Pictorial Album of the Navel Stores Industry Library and Archives, Forest History Society, Durham, NC, USA

Gum collection season ran from March through October, when the gum flowed freely. After that, gum that had hardened on the cut face of the tree as it ran into the cup, called scrape, was scraped off, packed in a specially designed box, then when full, dumped into resin barrels and hauled to the still. Trees were exhausted of gum production in 4-6 years, the trees were cut for wood products, and the camp moved to another location. The 8-10-foot-tall face was permeated with gum and was cut from the rest of the tree when it was harvested for wood until new technology enabled extracting the gum from this "cat face." These cat faces, which didn't easily rot due to the preservative effects of the gum, made excellent sources of rich resin-soaked firewood.

Dwellings for workers and their families were usually two room log huts with no glass windows, furnished with a few chairs, a bench, one or two beds, a corner cupboard, and an oven. Yards typically included a well with a cypress bucket or pitcher pump, a sweet potato house (earth-covered trench), and a vegetable garden with corn, sweet potatoes, peas, cabbage, and collards or turnips. A common smokehouse for preserving meat served several dwellings. Chickens scratched for food in the yards and were put in coops at night. Hogs and milk cows foraged for themselves in the forests. The general manager's house was larger with more rooms and furnishings, glass windows, and outbuildings that included a kitchen, smokehouse, and storage house for canned goods. Food supplies were supplemented with such game as the men could shoot while working the trees, usually squirrels, rabbits, opossums, and an occasional deer or turkey. The women and children caught catfish, bream, perch, and sometimes bass in the streams and ponds in the woods around them.

Pine gum unavoidably got on the hands and clothing of laborers. Raw gum is extremely sticky and difficult to clean off clothes and skin. While its adhesion to clothing was only a nuisance, its contact with skin could cause dermatitis. Treatment of this skin irritation is ineffective until the offending agent is removed. Laborers who found themselves afflicted during the harvest season had to wait until November at the earliest for a cure.

The wilderness nature of the turpentine forests contributed further to harsh working conditions. Black bears, alligators, poisonous snakes, malarial mosquitos, ticks, and chiggers found in the woods could make turpentine work a miserable and sometimes hazardous occupation. Florida's heat and humidity, particularly during the summer, were difficult to tolerate.

It was a primitive, hard, isolated, migratory and unique way of life. But for those who endured it, it developed immense character and fortitude. Many of the citrus groves you now see in Florida are planted on land that once grew pine trees. Turpentine operations, with the logging that followed, cleared this land, and it was subsequently planted in citrus groves.

The Florida Timber Industry

History Timeline

1830s – 1850s

Joseph Forsyth and Andrew and Ezekiel Simpson (Forsyth and Simpson Arcadia mill) built one of the earliest sawmills in west Florida. Located near Pensacola, this water-powered mill was producing over 250,000 board feet of lumber per year (an average of seven hundred board feet per day) by the mid-1830s. By the end of the decade the mill sent out an average of over 900,000 board feet of lumber per year.Efficiency and production were improved in the north Florida mills with the introduction of steam engines and circular saws. The lumbering industry in north Florida was becoming well established. Most of the steam-powered mills in north Florida were similar: small, family-owned mills with a roof covering the saw, steam engine and furnace which employed from six to eight laborers and produced about 300,000 to 450,000 board feet of lumber per year (about 1,300 board feet per day). These small mills were run by the owner and his family, with seasonal help being hired during peak production periods. Since hauling logs any distance was costly for the small mills in the interior of north Florida, these mills mainly served the local area.

Homestead Act of 1862 and Southern Homestead Act of 1866

The Act of 1862 granted adult heads of families 160 acres of surveyed public land if they improved and lived on the land for five continuous years. A "tree claim" could also be made if ten acres of timber was planted and successfully cultivated. The Act of 1866 encouraged whites and freed slaves to own farms of either 80 or 160 acres, the price of which was $5.00- for the farm, not for an acre. Most timberland at the time was selling for $1.25 an acre.

1867 – 1876

Leaving from Chicago, John Muir begins his long walk to Florida on September 5, 1867. He describes the extent of the longleaf pine forests in great detail in his writings. The first lumber census reported a mill tally for the South of 1.6 billion board feet for the year. By 1879 it had almost doubled. The beginning of Franklin County's timber industry starts with several small lumber mills. Timber supplies seemed endless, and the industry began to grow. The timber industry would support a large portion of the area's economy for many decades.

Despite the fact that loggers and sawmill workers in north Florida worked long hours with low pay, little opportunity for advancement and a high risk of serious injury, labor walkouts and strikes were not common and the ones that did occur were unorganized and uncoordinated. Sawmill workers attempted to strike at a Jacksonville mill in 1873, but after a month the strike was recognized as a failure. Some of the workers returned at pre-strike rates, others were replaced and left jobless.

Due to hard times brought on by the Civil War, most southerners did not have the capital to purchase timberland. Timber speculators from the north came by the trainload to make their purchases. Under the Timber and Stone Act of 1878 transfer of public lands to private ownership did not require the property to be used for agriculture. Abuse resulted in soil erosion and watershed destruction which led to its repeal, many years later, in 1955.

1880 – 1888

The population of Florida had doubled since the end of the Civil War. Thousands of immigrants came to this still largely undeveloped area looking for employment or land, many were subsistence farmers or field hands from southern Georgia, southern Alabama, Mississippi and the Carolinas. Mill fires were reported weekly in the state's newspapers. These fires could shut down an entire plant, causing lost wages for the workers and affecting the operator's bottom line. Companies began posting "No Smoking" signs within their mills, but most were ignored. One account from the March 20, 1880 diary entry of an Apalachicola lumber mill superintendent by the name of Samuel Floyd read "Posted a sign in the mill according to instructions, "No Smoking", I had a lighted cigar in my mouth at the time."

Lumber grading was first adopted by southern mills as a way to standardize the quality of the boards produced. The Knights of Labor, an organization for skilled lumber workers, were making successful inroads into organizing the north Florida lumber industry. Membership increased from 104,066 to 702,924 between 1885 and 1886. Outbreaks of strikes occurred as organizers moved into the region, some of which resulted in temporary gains for the lumber workers. Local assemblies were organized in several north Florida counties including Franklin and Jackson, yet by 1893 all of them ceased to exist. Few details of the social activities or labor organizing activities of these assemblies were recorded. Internal feuding tore apart the Knights in Florida from the mid-1880s to the early 1890s and by 1892 the General Executive Board of the Knights of Labor revoked the state charter. The Knights regrouped in northwest Florida for a short time but disintegrated after 1902.

In 1888, Chicago investors had purchased 122,000 acres in Florida for as little as 25 cents per acre. Tracts in other southern states were transferred for about the same price. By 1892, yellow pine and southern bald cypress imported to the North totaled a hundred million feet annually.

By the late 1890s, Part-time seasonal workers were becoming a thing of the past. North Florida sawmills were now employing more permanent workers, and by 1910 many of them had never known any other occupation.

1900 – 1917

A "car house" rented for four dollars a month. These crude, portable homes were made from remodeled boxcars that were lifted from the tram tracks, set down a few feet from the right of way and fitted with windows for the loggers and their families. Tents and cabins were also used for housing in the camp towns, the company store, physician's office and other services were housed in these portable quarters as well.

Turpentine production peaked. The cross-cut saw was the latest in modern tools, making its appearance at the turn of the century. The eight-wheeled logging wagon was replacing the carts and wagons used in earlier days. These more efficient wagons were well suited for the north Florida terrain and could carry more logs. Oxen were used to haul the heavy loads on a long trip and mules for the shorter ones.

1909 was a peak period for the exploitation of the southern forest. More than 16 billion board feet of lumber was produced in those twelve months (45% of the total U.S. harvest).

With northern timber companies buying up private and public lands, poor blacks and whites were left unable to purchase land. Work in the forests offered better opportunities than the agricultural jobs that many southern men traditionally held, so despite the dangerous conditions, low pay and long hours, many of these men were giving up farm work and making a career of lumbering. The pay rate for laborers in north Florida sawmills was 10.5 cents per hour. By 1907 it had increased to 13.67 cents per hour, still far less than their northern counterparts were making.

1917 – 1924

Worker's Compensation and the various problems confronting labor were high priorities in the newly elected administration of Florida Governor Sydney Catts. Numbers of injured workers had attempted to sue their employers for work related injuries, but few received any type of settlement. Harsh resistance from businessmen in the lumber industry delayed the passage of workmen compensation legislation until the 1930s.

In 1918 the Southern Pine Association began its "Safety First" campaign due the huge number of mill accidents that were leaving the workers severely injured, maimed and in many cases dead.

By the early 1920s, most of the virgin timber in north Florida was gone and the naval stores industry was declining.

1924 – 1932

Clarke-McNary Act authorized the government to acquire those lands with the most potential for timber production and develop them as demonstration areas. The U.S. Forestry Service, along with the American Forestry Association began a campaign to educate the public on timberland management. The Dixie Crusaders, as they were known, traveled all over the southern backwoods with their exhibits. Edward Ball began buying large tracts of land for timbering with funds that he and brother-in-law Alfred I. du Pont had earned by investing in failed banks during the Depression. His initial purchase was 66,000 acres, but he ultimately acquired one million acres.

In 1932, a Georgia chemist by the name of Charles Herty developed a chemical process to make paper out of southern pine. As word spread about this new process, pulp mills began to spring up all over the south. The demand for pulpwood land increased and land prices began to rise.

Ornithologist Herbert Stoddard's publication titled The Bobwhite Quail, Its Habits, Preservation and Increase, which came out in 1931, helped convince the Forest Service that suppressing fires was harming the longleaf forests and was one of the reasons that the bobwhite quail populations were in decline. By the 1940's the Forest Service instituted what they called "prescribed" burning and began to apply fire management to forest lands.

Mid-1930s – 1940s

The vast virgin pine and cypress forests that had covered north Florida were nearly depleted. Reforestation was not a common practice, and if it was used, the timber could not be harvested for another twenty years.

In 1936, the St. Joe Paper Company was formed as part of Edward Ball's dream of a model community a year after Alfred du Pont's death. The paper mill began operations in 1938 in Port St. Joe, employing hundreds of workers. It was sold and shut down in 1999. In 1937 Paul Hursey was hired as one of the South's first professional foresters. His job responsibilities included fighting fires, marking timber for cutting, making management plans and laying out railroad trams. Foresters in state agencies conducted tours and gave demonstrations for the industrial leaders so that they could better manage their lands and increase their income.

1940s

Soil conservation districts were organized. Foresters in the Soil Conservation Service, as well as those in the state and Agricultural Extension Service, gave programs to landowners that sought expert advice on the management of their woodlands. They also educated the public throughout the region with their publications, programs, and demonstrations.

As the last great trees were cut, lumber production began to slow and then ceased altogether in many of the company towns. These once thriving company towns were dismantled and the company culture they spawned died with them.

Reforestation is the process of planting trees back into areas that had trees cleared or destroyed by logging or natural events. Although it began in the late 19th century, by the late 1940s it became widely practiced.

Today's Florida Timber Industry

Forest land is defined as land that is 10 percent forested by trees of any size. Timberland is forest land that is available to harvest and capable of productivity over a long period of time. In 2017, there were 16,970,464 acres of forest land in Florida, over 15 million of which is timberland. Most are privately owned; the rest is owned by state local, state, and federal governments. More than 5,000 items are made from forest products, ranging from furniture to pharmaceuticals. They can be found in your kitchen (fruits, nuts, spices, artificial flavorings, food additives, wine corks, sports drinks, root beer, dish washing liquids); your medicine cabinet (soaps, cough syrups, cough drops, shampoo, aspirin, adhesive bandages, skin lotions, tablet coatings, toothbrushes, cologne, cosmetics, cancer-fighting drug Taxol); your playroom (crayons, football helmets, computer casings, sporting equipment, musical instruments); your garage, garden and laundry room (car wax, tools, paints, charcoal, garden mulch, cleaners and laundry detergents); and throughout your home (carpet and fabrics).

Forests have a positive impact on your life outside your home as well. Environmentally, they provide habitat for wildlife; act as natural air and water filters; and offer many recreational opportunities and much-needed green space. Financially, the forestry industry employs more than 124,000 Floridians and infuses $25 billion into the state's economy.

Timber operations include growing and harvesting timber on a 10-year or longer production cycle, or gathering of forest products such as gums, barks, balsam needles, rhizomes, fibers, Spanish moss, ginseng, and truffles. These operations require natural forest or suitable areas of land that are available for a long duration. Maturation time for timber depends on the species of tree, climatic conditions, and the intended purpose of the timber. Timber harvesting typically requires specialized machinery that is unique to the industry.

Acres of Timberland by Forest Type Group

Forest Type Group	Thousand Acres of Timberland
Longleaf–Slash Pine	5,599.5
Oak-Gum-Cypress	3,748.3
Oak-Hickory	2,653.6
Loblolly-Shortleaf Pine	1,796.9
Oak-Pine	1, 539.2
Total	15,337.5

Source: USDA Forest Service, 2020. Forests of Florida, 2017. Asheville, North Carolina.

Industrial timber products include pulpwood, saw logs, veneer logs, composite products, and roundwood harvested for uses such as poles, posts, firewood, logs for log homes, and mulch. Pulpwood is timber harvested for making wood pulp for paper production. Pulpwood is of primary importance when thinning a stand. It is typically a small tree measuring 6-9" diameter at the harvester's breast height (DBH). Saw logs are logs of suitable size for sawing into lumber processed at a *sawmill*. Trees cut for lumber fall into two categories, hardwood lumber and lumber from conifers. Lumber from hardwoods and pines typically is sawn from trees with diameters greater than 14" DBH. Trees are cut into lumber but some of the extra material is converted into chips for fuel or paper production. Sawtimber is measured in board feet or tons. There are 1,000 board feet in a ton. The value of these trees is heavily dependent on tree quality meaning straight, solid logs with little to no defect. Both hardwoods and softwoods are used for making pulpwood. Hardwoods include acacia, aspen, birch, eucalyptus, maple, pacific albus and rubber tree. Softwoods include pine and spruce.

Veneer logs are cut for peeled or sliced wood veneers and plywood. Trees in this product class have a diameter size of 16" or more. By means of a large lathe, the tree is converted into continuous sheets of thin wood. This is used in the manufacture of plywood and furniture, depending on the type of tree. Veneer and plywood is measured in tons or board feet. Value is heavily dependent on tree quality. North America's earliest composite wood manufacturers built their businesses by making high-quality wood-based products from sawmill waste, such as sawdust, shavings and wood chips—recycling waste material that would otherwise be landfilled or burned. Today's manufacturers take these original practices further by incorporating post-consumer wood waste and utilizing waste from sawmills that source wood from sustainable, locally managed forests. Products include fiberboard, particleboard, hardboard, engineered wood siding/trim, and decorative surfaces.

Timber harvesting methods include clearcutting, the shelterwood method, and the seed tree method. Clearcutting reproduces a new even-aged stand by completely removing the mature stand. The shelterwood method involves the removal of most of the mature stand at the end of the rotation, but a portion of the mature stand is left standing. The shelterwood method serves three basic purposes: To prepare the stand for production of abundant seed, to modify the environment in a way that promotes germination and survival of the selected species, and to build up the amount and size of advance regeneration to ensure the prompt restocking of the new stand following overstory removal. The seed tree method is similar to the *shelterwood method* in that it entails removing most of the mature overstory and leaving a portion standing. However, the **seed tree method** leaves only a few residuals as a seed source only. The residuals from this cut are too few and scattered to provide shelter. The residual trees of a seed tree cut must: be genetically superior trees of the selected species (*dominants*), be sturdy and windfirm, be able to survive exposure, flower and produce abundant seed, be the VERY BEST trees of the mature stand.

Prescribed Fire or Controlled Burning is controlled application of fire to vegetative fuels, under specified environmental conditions and following appropriate precautionary measures, to achieve specific objectives including hazard reduction such as risk of wildfires, control of undesirable species, wildlife habitat improvement, improved access and visibility, disease control, enhancement of appearance and aesthetics, and manipulation of depth, chemistry, and decomposition of the site. Although fire is a very effective and inexpensive land management tool, it can be disastrous under adverse conditions. Seek the assistance of a professional if you are not confident about your knowledge of prescribed fire. Fire is very useful to forest managers in many ways and must be applied at the time of year when there is a good chance of reaching the given objectives. The distinction between managed fire and wildfire must be made clear to the public so that land management policy will support the safety, health, and economic viability of Florida's public and private forests.

Tree harvester describes any number of machines that cut (fell), gather, and pile (bunch) trees. The machine either drives to the base of the tree or reaches to the tree with a boom (extending arm). It severs the tree, using either a circular saw or a chainsaw-type head which cuts the tree. After felling, trees are piled into hitches in or along skid trails for removal to the landing by a skidder. Mechanized operations tend to be whole-tree operations, which remove the entire tree from the forest for processing at the landing. This requires the support of several pieces of auxiliary equipment such as a loader, de-limber, slasher and chipper. The loader moves products around the landing as trees are processed into logs and chips. The de-limber removes branches, and the slasher cuts logs to length. Poor-quality stems and tops are run through the chipper to create biomass.

A *forwarder* moves large, felled logs to a roadside landing. It differs from a skidder by lifting logs at least two feet above the ground, which reduces soil impact. Forwarders usually work with machines in logging operations that require cut-to-length results. *Log loaders* sort and stack logs into piles. When the piles are of sufficient size, loaders move them onto transport trucks. There are three types of loaders: wheeled, capable of quick movement around the landing area; tracked, which are slower but more stable, and trailer mounted, which is the least expensive but must be pushed or pulled into place.

The author estimating the board feet of lumber in a crop of pine trees in FFA competition, 1967.

Courtesy University of Florida. Institute of Food and Agricultural Sciences (IFAS)

Courtesy Trimble Forestry

Courtesy Truck Loggers Association

The Florida Cattle Industry

The author cultivating corn he grew for cattle feed.

One of the author's corn crops.

The author feeding corn shucks to his cows.

The author grinding his corn into cattle feed from the crib he built.

The author, butchering a steer he pen-fed.

Cattle were my first love in agriculture. I earned college money buying and selling cattle, and I also had a herd of brood cows and a bull. For a few weeks during summers when I was a teenager, I worked cattle. It was for several families that together leased tens of thousands of acres of range land and ran cattle on it. The round ups consisted of bringing cattle out of the woods, prairies, and swamps to a main herd that was growing in size as cattle were brought to it.

This main herd was being driven down a dirt or rutted road to cattle pens by men and boys on horses. Once penned, the male calves were castrated, and each family's brand was placed on the calves

The author on his horse in 1968.

that were with the cows that had their brand. Castration is the removal of the testicles usually by surgical methods. There is also nonsurgical castration by injecting a chemical into the testicles, and by emasculation, which entails crushing the spermatic cords with a clamping device. Once castrated, the male calf is referred to as a steer. Beef from steers is preferred over beef from bulls because castration improves the color, texture, tenderness, and juiciness of the meat. And cow-calf producers and stocker operators benefit from castrating bulls because the market pays a premium for steers, partly because of consumer preference and partly because of their quieter disposition and ease of handling in the feed yard.

Some of the yearlings were loaded onto trucks and carried to the livestock market for sale. The land they leased was so far away that we stayed there and camped on it until the round up was finished. Our group was six or seven men and about as many teen-age boys, plus dogs to help herd the cows. For us boys with endless energy, twelve-hour days on horseback or branding calves were followed by nights of wild hog hunting or catfishing in the creek on the property. The men stayed in camp at night and played poker.

Overview

This section draws heavily on information from the Florida Department of Agriculture and Consumer Services.

Most people today believe that the cattle industry was first established in the American West, and over the years popular movies have supported that idea. But Florida is where cattle ranching began in America, and today it is a major cattle producer, and home to a ranch with the largest brood cow herd in the United States. Cattle were first introduced into North America in Florida in 1521 by Spanish explorer Juan Ponce de Leon. For nearly 500 years, the Florida cattle industry has contributed significantly to Florida's economy and natural resources. Nearly half of all Florida agricultural land is involved in cattle production. Florida cattle producers are good stewards of the land as owners and caretakers of thousands of acres of pristine native range and pasture land. In recognition of their conservation efforts, Florida has had several regional winners and one national winner of the National Cattlemen's Environmental Stewardship Award. Lands used for cattle production are also important "green space" for wildlife and native plant habitat, aquifer recharge, and carbon recovery. Grass and forages used for cattle production are renewable resources occurring in a great variety of landscapes typically found on cattle ranches, including improved pastures, wetlands and marsh, woodlands, and prairie. Biologists conclude that bird and wildlife populations thrive on lands used for cattle production. Florida's cattle industry was a leader in the formulation and adoption of agricultural industry "Water Quality, Best Management Practices" and other standards.

Florida is a cow-calf state. The primary cattle crop is calves, which are shipped to other states in order to grow, graze and/or be fed grain and be processed into beef. Florida's sandy soils do not produce

Photo by Carlton Ward, Jr. / CarltonWard.com

the high yields of corn and other grains used to feed cattle to produce high quality beef like other states such as Iowa and Kansas. However, Florida's mild climate is ideal for producing and raising calves and grass pastureland. The value of Florida's cattle breeding herds and calves is estimated at almost $1.3 billion, with annual calf sales of over $500 million.

History

This section draws heavily on *Florida Cattle Ranching* from the Florida Memory Project.

In Florida, those who own or work cattle traditionally have been called cowmen. In the late 1800s they were often called cow hunters, a reference to hunting for cattle scattered over the wooded rangelands during roundups. At times the terms cowman and Cracker have been used interchangeably because of similarities in their folk culture. Today the western term "cowboy" is used for those who work cattle.

Florida ranching has evolved from many different cultural traditions, though the most important sources were the marshy coastal areas of Andalusia, Spain, and the hill regions of Britain and Ireland. Florida's Andalusian cattle were the first in today's United States. They were brought to Florida by the expeditions of the Spanish explorers during the early 1500s and escaped and survived in the wild. Organized ranching began with the founding of St. Augustine in 1565, when cattle from Spain and Cuba formed the basis of herds that fed the garrison and surrounding communities. In addition to herds owned by the Spanish and Indians, wild cattle flourished in the rangelands and prairies. Eventually Spanish colonists began exporting cattle to Cuba. During the 1600s, Spanish clergy raised cattle at the missions, where many Native Americans learned to tend them.

By 1700, Florida contained approximately 34 ranches and 20,000 head of cattle. After British-Creek Indian raids in 1702 and 1704 devastated

Florida cattle ranchers, Indians sustained cattle raising in Florida. Many had established large herds of wild cattle and stock acquired from the Spanish. Cattle herding was vital culturally and economically to the early Seminoles, who remained Florida's major livestock producers throughout most of the 1700s. In early Florida, Europeans, Americans, and Indians stole cattle from each other. Rustling became particularly widespread by the second half of the 18th century and was one of the elements that led to the Seminole Wars.

When the U.S. took possession of Florida in 1821, it was described as a "vast, untamed wilderness, plentifully stocked with wild cattle." Florida "scrub" cattle were descended from the mix of Spanish and British breeds. These hardy creatures survived on native forage, tolerated severe heat, insect pests, and acquired immunity to many diseases. Early Florida cowmen survived in difficult conditions. They fought off panthers, wolves, bears, and cattle rustlers. Cowmen spent weeks or months on cattle drives across difficult marshes and dense scrub woods, often enduring burning heat, torrential thunderstorms, and hurricane winds. From central Florida they sometimes drove cattle as far as Jacksonville, Savannah, and Charleston. This gradually changed in the 1830s when the cattlemen re-established trade with Cuba, and Tampa, Punta Gorda, and Punta Rassa (near Fort Meyers) became important export ports.

The number of cattle increased rapidly from the 1840s until the Civil War. Florida was second only to Texas in per capita value of livestock in the South. After the Armed Occupation Act of 1842, cattlemen from the overstocked states of Georgia, Alabama, and the Carolinas homesteaded 200,000 acres in Florida. Some seized range territory that the Seminoles had been forced to relinquish because of the Seminole Wars. The newcomers often brought foundation herds that interbred with wild scrub cattle. Few cattlemen owned grazing land since there was extensive open range. By mid-century, ranchers were running large herds on the extensive open range in central and south Florida.

Wars provided an economic boost for Florida cattlemen, who provisioned armies during the Seminole, Civil, and Spanish American Wars. Although the Civil War disrupted the Cuban trade, Florida cowmen became beef suppliers to both armies. Hides, tallow, and meat from Florida were so important for the Confederacy that a Cow Cavalry was organized to protect herds from Union raiders.

After the Civil War there were still wild cattle in Florida, as well as attractive markets to the north and south. During the next three decades, trade boomed with Cuba, Key West, and Nassau, and Florida became the nation's leading cattle exporter. From 1868 to 1878, ranchers received millions of dollars in gold doubloons for over 1.6 million cattle exported to Cuba. The Cuban commerce provided income to cattlemen, merchants, and shippers, and contributed to the state's recovery from Reconstruction-era depression. After World War II, under the administration of Florida Governor Fuller Warren, open range was ended, and Florida cattle were confined to fenced lands.

Technique and Process

During the 19th century, particularly after the Civil War, wild cattle were plentiful in Florida. The technique to get them to market consisted of cowmen on horseback gathering individual and/or small groups of cows out of the woods and swamps and moving them to a larger main herd being driven down a trail to one of Florida's ports. As the cattle drive progressed, the main herd grew larger.

Roping wasn't practical in the Florida swamps and brush, so a saddle horn wasn't needed. Thus, McClellan army saddles were popular. Leather whips were used to control the direction of the cattle. They were typically 12 – 14 feet of braided leather with 14-inch wooden handles and buckskin poppers. A skilled cowman could make them crack like a rifle shot and was accurate enough with them to cut a leaf from a branch. Two techniques were/are used, one to move the herd and the other for moving a single cow in a specific direction. In the one to move the herd, the cowman twirls the whip around his head, then suddenly jerks his wrist downward and backward in the opposite direction, causing the whip to make a loud popping crack. The one for directing a single cow back into the herd or into a pen entails the cowman casting the whip straight in front of him, pulling it backward until his arm and the popper go behind him, then moving his arm upward as if to throw a baseball, and suddenly flicking his wrist in reverse motion and downward, causing a loud popping noise to motivate the cow to move, or if necessary, to strike the animal to make it move in the desired direction. The use of these loudly cracking whips to work cattle is where the term Florida Cracker originated.

Cowmen would often round up cattle, put them in pens, brand them with their mark, castrate the males they wanted to sell later, then turn them loose. At a later date, they would round up the cattle with their brand, brand calves following a cow with their brand, brand any unbranded cattle, and drive them to market. The markets were port towns from which the cattle were loaded aboard ships and exported.

Courtesy, U.S. Library of Congress

Today's Florida herds are on privately owned land, usually with pasture planted in Bahia grass or Bermuda grass. They are rounded up by cowboys on horseback and herded into cattle pens. These cowboys are independent contractors, and provide their own trucks, horses, herding dogs, whips, and other equipment. Once penned, the cattle are each put in squeeze chutes with head gates to hold them while they are ear tagged and/or branded, dewormed, and vaccinated against disease. Calves are castrated as soon after birth as possible because the younger they are, the less traumatic it is for them. Breeding bulls must meet high standards of genetics. Vastly improved breeding genetics proliferate in Florida's beef cow herds today. In addition, some cattle producers are turning to artificial insemination to advance genetic progress and reduce the expense of maintaining and managing breeding bulls.

Courtesy Powder River Manual Cattle Chutes

Beef calves typically nurse their mother for the first 6-7 months after they are born, until they weigh 400-500 pounds, when they are called feeder calves. Feeder calves are weaned and put on grass pasture, often with some supplemental grain feeding, for another 6-9 months, which is called back grounding. Back grounding enables the cattle to finish growing into adulthood and develop the body structure and maturity to be fed grain that enables them to produce high quality beef. At about 12 – 18 months old and a weight of about 700 – 900 pounds, the cattle are taken off of pasture and put in feedlots. These are large pens where cattle are fed grain daily consisting mostly of corn, but also milo, barley, and oats. They are also given hay for roughage. They remain in feedlots for 6 – 7 months where they gain another 400 – 600 pounds and will produce choice and prime grades of beef. From there, at weights from 1,100-1,500 pounds, they go to a slaughterhouse.

Marketing

Beef cattle are marketed through public auction markets, direct private sales, graded and pooled sales, tele-auctions, video auctions via satellite and the internet, and retained ownership. They are priced at competitive bids, and increasingly with forward contracts at agreed upon prices or with pricing mechanisms tied to the commodity cattle futures market. **Public auction markets** provide sellers with competitive bidding and prompt cash payment. Auction markets are open to all sellers and buyers and to the public. **Direct private sales** are from the rancher to the buyer. The seller controls the marketing process and can point out and promote positive aspects of their cattle. Buyers can see the total ranch program and recognize value-added production practices used on the ranch. **Graded and pooled sales** are the combination of small lots of livestock from several area livestock producers into larger, uniform lots of animals. The large numbers of cattle attract additional buying competition. **Tele-auctions** entail the use of a telephone conference call to facilitate an auction between buyers and sellers. Pictures of the cattle are displayed as still images or video. **Video auctions** broadcast over satellite and/or the internet with videos provided for advance viewing can reach a very large number of potential buyers and sellers. **Retained ownership** involves holding ownership of cattle through the next one or two stages of production. For example, a cow-calf producer would retain ownership of the cattle through the back grounding phase, to the feedlot, and maybe also through the feedlot phase as well. This enables the cattle producer to more fully capitalize on the value of good management practices and superior genetics that may not be fully apparent until the animal reaches the feed lot or slaughterhouse. However, the increased time between when the animal is a feeder calf and when it is ready for slaughter increases price risk and postpones receipt of revenue (Lacy, et.al.; Parish, et.al.).

The Florida Commercial Fishing Industry

Photo by Karl Holland, Courtesy State Archives of Florida, Florida Memory Project

My master's thesis at the University of Florida was about the contribution that Florida's commercial fishing industry made to the Florida economy. Doing the research for it required that I conduct interviews and spend time with numerous commercial fishermen. During that process, I came to appreciate the risks, hardships, and rewards of commercial fishing as a way of earning a living and as a way of life.

Fishing was one of the first activities to become an integral component of Florida's economy. Prior to the arrival of the earliest pioneers, Florida's native Americans were heavily dependent on fishing for their daily sustenance. It was the vast abundance of fresh readily available fish and shellfish in numerous bays and lagoons along Florida's meandering peninsula and island coastline that insured its first settlers against starvation and also provided a basis for trade with inland colonists. As the population increased, more people turned to fishing to supply an increasing demand for marine products. The result was the birth of Florida's commercial fishing industry. Commercial fishing in Florida is geographically divided into two principal areas, the East Coast and the West Coast. The East Coast fisheries are part of the South Atlantic fishery region of the United States. This region is comprised of the Atlantic Ocean off the coasts of North Carolina, South Carolina, Georgia and the Florida East Coast. The West Coast fishery is in the Gulf fishery region which includes Texas, Louisiana, Mississippi, and Alabama. Florida is the only state participating in two regional fisheries (Gulf and South Atlantic), which, when combined, is considered the "Southeastern" fishery region of the United States.

Florida ranked 11th among U.S. states in 2021 for fresh seafood production with 110 million pounds harvested and a dockside value of $262 million, which is 4.2 percent of U.S. total value. The leading species by value was shrimp, at $63.7 million.

Florida's commercial fishing industry is comprised of four sectors: fishing, wholesaling, processing, and retailing. Fishermen combine management, labor, and capital to obtain raw marine products from the ocean. Seafood wholesale markets and fish dealers purchase the fishermen's catch at dockside, prepare it for sale in fresh form, and sell it wholesale to supermarkets and restaurants, and sometimes retail to consumers. Seafood processors produce smoked, breaded, canned, frozen, and stuffed fish and shellfish. They purchase fish and shellfish from fishermen, wholesale fish dealers, terminal markets, and other available sources. The retailing sector consists of seafood markets, supermarkets selling seafood, and restaurants. The associated economic activity of these four sectors results in an economic contribution to Florida's economy by the Florida commercial fishing industry of $1.2 billion and the creation of 18,000 jobs (Hodges, et.al.)[1]

Nineteen species of shellfish and 74 species of finfish are landed in Florida. The most valuable twelve of these in order of value are shrimp, stone crab claws, grouper, spiny lobster, blue crab, oysters, king mackerel, mullet, swordfish, tuna, and Spanish mackerel. Florida fishermen catch 100 percent of the spiny lobsters, 97 percent of the stone crabs and over 84 percent of the grouper, pompano, mullet and Spanish mackerel landed in the U.S.

Unlike northern lobsters, Florida lobsters, also called Caribbean lobsters, have spines instead of claws, and thus are called spiny lobsters. These are two forward-projecting whip-like antennae that are about two-thirds as long as the lobster's body, covered in 1-2mm sharp spines that help protect them from predators. In addition to Florida, spiny lobsters are caught (trapped) in the Bahamas, Cuba, Honduras, Nicaragua, and Brazil. Spiny lobsters are harvested commercially in baited traps marked with buoys. Recreational harvest, which accounts for about 10 percent of Florida's lobster catch, is by diving using gloves, a lobster net, tail snare, and tickle stick to coax lobsters out of holes.

Expansion of the Bahama's offshore sovereign fishing grounds from three to twelve miles in 1969, and complete closure of the

---

[1] Although completed in 2002, this is the most recent comprehensive study of economic impact.

Spiny Lobster, Courtesy Florida Sea Grant

Bahama's continental shelf from foreign fishermen in 1976 contributed to a 50 percent reduction of spiny lobster landings in Florida between 1969 and the late 1970s. However, lobster management plans enacted in the early 1980s that placed size limits on lobsters harvested, provided a closed season, gear restrictions that prevented the use of spears and hooks, and measures to prevent poaching restored the lobster harvest back to the levels it was in the 1960s. The commercial and recreational harvest season for spiny lobsters in Florida is from August 6 through March 31. There's also a special two-day season in July for recreational harvest before the start of the general open season. Personally, I think spiny lobsters are sweeter and more flavorful than northern lobsters.

Florida stone crab season is from mid-October to mid-May. They are harvested in baited traps marked with buoys. Only the claws are taken, the crab is returned to the ocean, where it will regrow the claw or claws removed within a year. Stone crab claws are cooked immediately upon harvest, usually dockside. This prevents the meat from sticking to the shell. They are then sold to seafood markets and restaurants, where they are stored chilled or frozen. They can be eaten cold or hot and are best served with melted butter or mustard sauce. I put stone crab claws in a pot of cold water directly from the faucet and heat them until they just begin to boil. I crack them on a cutting board with a wooden mallet. Florida stone crabs are my favorite seafood.

Three techniques are used by commercial fishermen to catch fin-fish: vertical hook and line, longline, and trawls. Vertical hook and line is used for reef fish, which are fish that feed on plant and animal marine life on coral reefs. The main species of reef fish caught commercially are grouper, red snapper, amberjack and gray triggerfish. These fish are usually caught at depths of 75 – 300 feet, but often deeper. The fishing gear consists of an electric or hydraulic reel mounted on the gunwale of the boat with a metal arm extending over the side. This is known as a bandit rig and gets its name from the resemblance to slot machines found in casinos. Vessels typically have from three to six bandit reels on board. A typical bandit reel consists of a fiberglass reel that holds about 1,000 feet of stainless-steel cable; and L-bar or spreader, which keeps the leader from tangling with the main line; a pulley to feed the cable from the reel through the L-bar; a fiberglass arm; and an electric or hydraulic reel motor. Fishermen typically use circle hooks baited with squid or cut fish, with multiple short (about six inches long) lines attached to the main line, and hooks on each short line. The hooks are baited and lowered into the water. Once fish are on the short lines, the main line is reeled in, and the fish are removed and put on ice.

Stone Crab, Courtesy Longboat Key News

Longline fishing is commonly used for swordfish, tuna, shark, halibut, and sablefish, although other species are caught with this technique. The gear consists of a long line usually made of monofilament, called the main line, with baited hooks attached at intervals by means of branch lines called snoods. A snood is a short length of line attached to the main line using a clip or swivel, with a hook at the other end that is baited with squid or cut fish. The length of the line ranges from a mile to as long as 60 miles, and is buoyed by Styrofoam or plastic floats. Hundreds or even thousands of baited hooks can hang from a longline.

Bandit Rig, Courtesy Florida Sea Grant

Monofilament nylon gillnets, which are nets with a mesh size that allows the fish's head through, but catches on its gills or body, were used to catch schooling fish in relatively shallow water. These were primarily mullet, spotted sea trout, sheepshead, and bluefish caught close to shore in bays and inlets, and Spanish mackerel, caught further out at sea, usually 8-12 miles. Over 90 percent of the Spanish mackerel landed in the U.S. are caught in South Florida, primarily in the Fort Pierce – Port Salerno area from December through February. A small airplane, called a spotter plane, was often used to help locate the schools of fish. However, there was concern among recreational fishermen and environmentalists that the use of gillnets was leading to overharvest of inshore fish stocks – mullet, redfish, Snook, spotted sea trout, and others. Consequently, Amendment Three of the Florida Constitution, known as the net ban, was voted into law in November of 1994. It made the use of gillnets in Florida waters unlawful and limited the use of other types of nets to a maximum of 500 square feet. This rule virtually wiped out the near-shore net industry – overnight. Since the gillnet ban, mullet and other inshore species are primarily harvested with hand-thrown cast nets, which are not nearly as productive as gill nets. The net ban has had a negative impact on commercial fishermen, seafood wholesalers and retailers, marine supply dealers, and consumers of locally caught seafood.

Various types of gear are used to capture shrimp, including cast nets, haul seines, skimmer nets, wing nets, traps, and beam trawls. However, the otter trawl, with various modifications, is the dominant gear used for offshore shrimp harvesting. Since the 1950s, double-rig trawlers have replaced single otter trawls. A double-rig trawler consists of two nets each about 40-50 feet wide made of nylon or polyethylene webbing suspended from two outriggers located on the port and starboard sides of the vessel. Each net consists of a heavy mesh bag with wings on each side designed to funnel the shrimp into the net. A pair of trawl doors positioned at the end of each wing holds the mouth of the net open by exerting a downward and outward force at towing speed, which is about 2.5 – 4.0 knots (3 – 4.5 miles per hour). The outriggers are metal poles typically 24 feet long and four inches in diameter, although their size varies. To prevent fowling of the two trawl nets, one trawl is set out and fished about 150 feet farther astern than the other one. Usually, the starboard trawl is given the longer scope and set out first. When the second trawl is set out and clear, both trawls are lowered simultaneously to the desired depth. At the end of a drag, usually about three hours in duration, the trawls are hauled in simultaneously. The trawl closest to the vessel comes in first, it is pulled over the boat using a long-handled pole hook, and the shrimp are dumped from it. Then the second trawl net is retrieved.

In the United States, sea turtles are protected under the Endangered Species Act. Turtles are air-breathing reptiles. When they are caught underwater in nets or on lines, they drown if they are unable to reach the surface for air. Concerns over the by-catch of sea turtles by shrimp trawls led to federal law that requires the use of turtle extruder devices (TEDs) in the throat of shrimp trawls utilized in state waters. TEDs are two-dimensional net inserts with large openings that enable turtles to escape. Also, environmental concerns over the finfish by-catch associated with shrimp trawls led to the requirement that by-catch reduction devices (BRDs), which enable species other than shrimp to escape, be placed in shrimp trawls beginning in 1998 (Adams).

The Magnuson-Stevens Fishery Conservation and Management Act of 1976 (FCMA) extended the U.S. jurisdiction of fisheries from 12 to 200 miles and created eight regional fishery management councils responsible for the fisheries that require conservation and management in their region. The councils are comprised of both voting and non-voting members representing the commercial and recreational fishing sectors in addition to environmental, academic, and government interests. The key objectives of the FCMA are to prevent overfishing, rebuild overfished stocks, increase long-term economic and social benefits of the fisheries, and ensure a safe and sustainable supply of seafood.

Florida participates in two fishery management councils: the Gulf of Mexico and the South Atlantic. The Gulf council, based in Tampa, meets five times a year and also includes Alabama, Mississippi, Louisiana, and Texas. The South Atlantic council, headquartered in Charleston, South Carolina, meets four times each year, and in addition to Florida, includes North Carolina, South Carolina and Georgia.

The Florida Citrus Industry

The photos displayed in this section are courtesy of UltimateCitrus.com unless specified otherwise.

History

Christopher Columbus brought the first oranges to the New World on his second voyage in 1493, and early Spanish settlers planted orange trees around St. Augustine, Florida in the mid-16th Century. Florida's sandy soil and sub-tropical climate proved ideal for growing citrus. By the 19th century, citrus trees grew wild through many of Florida's forests, and groves were cultivated along major Florida rivers. Florida's annual commercial citrus production grew to five million boxes by 1893. Improved transportation opened new markets in the northeastern United States, and by 1915 production reached 10 million boxes.

The discovery of how to make frozen concentrated orange juice (FCOJ) in the late 1940's by the U.S. Dept. of Agriculture and the Florida Dept. of Citrus provided juice that tasted almost as good as fresh squeezed but was more convenient to make. Between 1945 and 1960, Florida orange production more than doubled as growers tried to keep up with the rapidly growing demand created by the new frozen concentrated orange juice (FCOJ).

A devastating freeze in 1962 cut Florida orange production in half, which was the largest reduction in fruit production since the discovery of frozen concentrate. This was a shock to the increasingly market-oriented citrus processing sector as orange juice supplies became inadequate to meet market demand and prices skyrocketed. It soon became evident that it would take years to restore damaged tree inventories.[2] Major Florida processors went in search of foreign sources of orange juice supplies and found the potential to develop them in Brazil. Brazil had a thriving orange industry that produced for the fresh market, a climate that rarely experienced freezes and that produced a juicy orange. Meanwhile, the greatly increased prices led Florida growers to begin planting trees at an unprecedented rate. And the Florida citrus industry of today was born, an industry that, until a few years ago, was second only to Brazil as an orange juice producer.[3] An industry that today, according to a study conducted by the University of Florida in May 2021, contributes $6.762 billion to Florida's economy and supports 33,300 jobs. And an industry that today is struggling to survive. Let's look at the success and failures, the opportunities and threats that have led the Florida citrus industry to the brink of disaster.

Citrus Production and Processing

Primarily citrus greening disease, but other events described in the last part of this chapter about Florida citrus titled "Recent Threats and Challenges", has reduced Florida's orange production from 242 million 90-pound boxes in 2003/04 to 41 million boxes in 2021/22. Florida's orange production for the 2022/23 season was forecasted to be 28 million boxes by the USDA on October 12, 2022. However, the impact of Hurricane Ian, a Category IV hurricane with sustained winds of 155 MPH that struck Southwest

2 Citrus trees are four years old before they produce a commercially harvestable crop, and their production increases until they are about 12 years old, when production levels out. A citrus tree will produce fruit for over 30 years if it does not succumb to diseases or a freeze.

3 Mexico is currently the world's second largest orange juice producer.

Florida on September 28 near Port Charlotte in Lee County and exited Florida over Merritt Island, resulted in an orange production forecast on March 8, 2023, of 16.1 million boxes for the 2022/23 season.

Because recent events make the current Florida citrus industry a-typical, the Florida citrus industry described in this paragraph will be as of the 2013/14 season, which more fully describes the industry as it has functioned for most of the last 40 years. Florida's principal citrus fruits are oranges, grapefruit, and tangerines. But oranges are by far its largest and most valuable citrus crop, accounting for 85 percent of production and 86 percent of the on-tree value of Florida citrus. Oranges are grown in Florida, California, Arizona, and Texas, but Florida accounts for 70 percent of U.S. orange production. California is second with 29 percent of production. However, California is the predominant fresh orange producer in the U.S., accounting for 85 percent of production. Florida's dominance as an orange producer is based on its dominance as an orange processor. Florida's climate produces a much juicer orange than California, which primarily produces oranges for fresh consumption. Consequently, Florida juices 95 percent of its oranges compared to California's 20-22 percent, and Florida accounts for 94 percent of domestic U.S. orange juice production.

Oranges are harvested for processing in Florida from November to June. They are hand-picked in the field, loaded into semi-trailers, each holding about 500 ninety-pound boxes,[4] and delivered to the processing plant. Oranges do not ripen after picking, and the deterioration process begins as soon as they are separated from the tree. At the plant, the fruit is dumped from the trailers onto conveyors, where it is carried through a washing process. The fruit is then

[4] In Florida, a box of oranges is a standard unit of measurement and weighs 90 pounds. It contains 220-250 oranges and produces 5.6 – 6.3 gallons of juice. Citrus fruit is now handled in bulk bins and trailers, thus the term "box" is a statistical measure only.

graded, and oranges not acceptable for processing are separated and diverted to be used for by-products. The Florida Department of Agriculture and Consumer Services (FDACS) under the authority of the United States Department of Agriculture (USDA) then tests for juice and solids content and calculates the unit of payment for growers, which is pounds of solids.[5] Next the oranges are diverted to storage bins labeled according to juice specifications represented by each tested load of fruit. Oranges are then selected from labeled bins to enable blending juice of optimal quality and transported via conveyer to juicing machines (extractors). After juice extraction, the stream of pulpy juice goes through a finisher (screen) where the pulp and seeds are removed, and along with the peel, are diverted to be used for by-products. At this stage, the juice is made into one of two product forms: frozen concentrated orange juice (FCOJ) or not- from-concentrate (NFC). Juice made into bulk FCOJ is sent to an evaporator where vacuum and heat are used to remove excess water to obtain a base concentrate of 65°Brix (65 percent fruit sugar), which is a six-to-one strength ratio to normal single-strength juice. The bulk FCOJ is then stored at 20°F (Fahrenheit) or lower in a tank farm or in 55-gallon drums until it is sold or packaged for sale. Juice made into NFC is either pasteurized, chilled, and packaged, or stored for future sale and/or packaging. NFC is stored in bulk form in two ways: frozen in 55-gallon drums, or pasteurized and chilled in million-gallon stainless steel, aseptic tanks.

When orange juice is evaporated into concentrate, the citrus essences and oils that give OJ its flavor are lost in the steam from the evaporator. At that point the OJ tastes like fruit sugar and water. Until the 1960s, this problem was solved by adding fresh single strength OJ back to the concentrate, a process called

---

[5] Pounds of solids are a measure of the soluble sugar solids in juice. Fruit and bulk juice bought and sold by processors is priced per pound of solids. One gallon of single strength ready-to-serve orange juice, such as consumers buy in a store, contains 1.029 pounds of solids. Oranges generally produce from 5.8 to 6.5 pounds of solids per 90-pound field box.

Harvesting Florida Oranges

Dumping Bag of Fruit into a Tub

cutback, which restored the lost flavor essences and oils, and reduced the brix from 65 to 42. But in 1960, scientists designed a device that attaches to the evaporator and captures these oils and essences. These compounds can then be formulated into "flavor packs" which are added back into the juice just before packaging. Nothing that was not naturally present in the orange is introduced; all the added compounds were present in the juice in the first place. Since the composition of the flavor packs can be adjusted, producers can ensure that the processed orange juice always tastes the same, even though the original juice may vary in flavor depending on the source of oranges and the time of the year they were harvested.

If NFC were just packaged and sold, it would quickly spoil. If there is to be an acceptable shelf life for NFC, the juice has to be pasteurized to destroy microorganisms and stop enzymatic activity. Moreover, consumers want juice year-round, but oranges do not grow year-round. And in Canada, England, and a number of other places, they don't grow at all. So the juice has to be stored, often for months. Juice processing saturates juice with air. Oxygen creates oxidative off flavors in stored juice unless it is frozen. To prevent these off-flavors, oxygen has to be removed from the air-space above the stored juice. However, removing oxygen strips the flavor essences and oils out of the juice like evaporation does, unbalancing the orange flavor profile. Thus, flavor packs of citrus essences and oils must be added back to juice stored aseptically to restore the orange flavor the same as with concentrate.

FCOJ is packaged by orange juice marketers into either frozen concentrated orange juice or chilled reconstituted (Recon) ready-to-serve (RTS) orange juice. Packaged FCOJ is made by adding single-strength juice or water and flavor oils and essences to bulk FCOJ to reduce it from 65°Brix to 42°Brix, which is a four-to-one strength ratio to normal single-strength juice. Then the product is packaged into 6, 12, or 24-ounce cans and sold in frozen form. To convert FCOJ into ready-to-drink orange juice, it is thawed and then mixed using the ratio of one-part FCOJ to three parts water.

Recon RTS is made by adding water and flavor oils and essences to bulk FCOJ to reduce it from 65°Brix to 11.8°Brix and then pasteurizing it. It is then packaged in cardboard cartons or glass containers and sold as chilled RTS orange juice.

A unique characteristic of both orange and grapefruit juice is that consumers can and do squeeze fresh oranges and grapefruit at home when they are in season. In contrast, other fruit juices such as apple, grape or pineapple are not as easy to squeeze or make in fresh form at home as citrus juices. Thus, most consumers have no measure of fresh-squeezed taste for these other juices. Consequently, fresh-squeezed sets a taste standard for orange and grapefruit juice that most other fruit juices do not have.

Finally, as women have taken jobs outside the home and real disposable incomes have grown, consumers have increasingly demonstrated a preference for and placed a value on convenience. Citrus juices then, which deteriorate rapidly in fresh-squeezed form, must live up to the higher flavor standard of fresh-squeezed, yet also meet consumers' growing preference for convenience. The challenges faced by citrus juice marketers to meet these exacting and often conflicting criteria tell the story of the successes and failures of the Florida citrus industry.

Citrus Juice Brands and Markets

Prior to the development of FCOJ in 1947 by the U.S. Department of Agriculture and the Florida Department of Citrus, the only orange juice that was available was canned or fresh squeezed. Canned orange juice had an inferior flavor and fresh squeezed was excessively time consuming to prepare. Thus, the U.S. orange juice market was small.

Dumping Tub of Fruit into the Body of a Fruit Loader Called a Goat

Goat Dumping Oranges Into A Semi-Trailer

Florida Citrus Processing Plant, Courtesy Marcos Valim, Citrosuco

Florida Foods Corporation, which later became Minute Maid, was instrumental in the marketing efforts that led to the successful introduction of FCOJ into the marketplace. Once in distribution it became apparent that FCOJ was the most economical way to bring the product of the grove and processing plant to the consumer's table, since concentration reduced packing, storage, transportation, and handling costs throughout the marketing chain. Regulations passed by the Florida Citrus Commission[6] maintained and improved the quality of FCOJ, and a generic advertising program funded by Florida growers and managed by the Florida Department of Citrus (FDOC) under the guidance of the Florida Citrus Commission helped to speed demand growth for Florida orange juice. Since FCOJ's quality was maintained and its price was reasonable, it soon became a leading seller among food products.

---

6 The Florida Citrus Commission, formed in 1935, is a 9-member board appointed by the Governor to oversee the Florida Department of Citrus (FDOC). The FDOC provides marketing, research, and regulatory support for Florida citrus products. Although it is a state agency, the FDOC operates under a state marketing order which establishes a means to raise revenue through a tax. Growers fund the FDOC through an excise tax placed on each box of citrus that moves through processing plants and fresh fruit packinghouses.

Minute Maid's success in marketing FCOJ caught the interest of a major veteran beverage marketer, Coca-Cola, who entered the citrus industry in 1960 with its purchase of Minute Maid. During this time another citrus processor, Tropicana, was able to provide orange juice year around by freezing single-strength juice and storing it for thawing, pasteurizing, packaging, and distribution when Florida oranges were not in season.

Oranges Being Washed Before Juicing

Oranges Being Graded Before Juicing

Juice Extraction

Anthony Rossi, Tropicana's founder, believed his customers, who were also milk customers on dairy routes, preferred ready-to-serve chilled orange juice and he could please them more during the off-season (July-November) with orange juice from frozen single-strength juice rather than from reconstituting FCOJ. Rossi felt that pasteurization, which did not heat orange juice as much as evaporation did, resulted in a better taste for off-season juice than juice that had been evaporated, and Tropicana's chilled orange juice was in ready-to-serve form, not requiring the preparation time of FCOJ. However, concentrate enjoyed a substantial cost savings in storage and distribution over single-strength juice, its preparation time was not excessive, and many consumers could hardly tell any taste difference between juice made from FCOJ and juice from oranges they squeezed at home. By 1969/70, FCOJ was still 78 percent of the U.S. orange juice market.

During the 1970's, stimulated by strong, growing orange juice markets, world orange juice supplies (from Florida and an emerging Brazilian citrus industry) doubled, which enabled the U.S. market to grow by almost 7 percent per year. Most of the orange juice advertising was the FDOC generic advertising program.

A development hardly noticeable by tracking only FCOJ volume since it had continued to grow, was that ready-to-serve chilled (RTS) orange juice had grown from 15 percent of the U.S. market in 1969/70 to 34 percent by 1979/80, while FCOJ share of market had shrunk from 78 percent to 62 percent. RTS had achieved its growth despite being higher priced to consumers than FCOJ. Canned orange juice, the only form of processed orange juice available year around before the development of FCOJ had shrunk to 4 percent of the U.S. orange juice market by 1979/80.

In the 1970's, a shift from glass to paper packaging facilitated reprocessing of bulk FCOJ into RTS reconstituted (recon) orange juice in locations close to major population centers. This was especially suited to dairies, which could easily switch from packaging milk in cartons to packaging orange juice in cartons. While these changes did not make RTS prices as low as FCOJ prices, it began to narrow this price gap and encouraged increased competition with FCOJ.

A historical evaluation of the U.S. orange juice market as of 1980 showed a market that had continued to demonstrate very strong growth and a market where the two leading brands, Minute Maid and Tropicana, together held only about a 33 percent share, the rest being held by private label (store brands), packer labels and regional brands. Proctor and Gamble (P&G) evaluated this situation and determined that there was an opportunity for a third major branded marketer. P&G purchased agribusiness baron Ben Hill Griffin's citrus processing plant in Frostproof, Florida in the early 1980's and began producing and marketing both FCOJ and RTS orange juice under the Citrus Hill brand. Rather than start regionally and grow slowly to a national brand, P&G positioned the launch of its citrus juices to quickly become a nationally advertised and distributed brand.

P&G's entry into the citrus industry as a branded juice marketer resulted in significant responses by both Minute Maid, by then called Coca-Cola Foods, and Tropicana. Both firms began to increase their market-focused approach to the orange juice business. Brand advertising expenditures grew dramatically, greatly outpacing FDOC's generic advertising and almost tripling between 1980 and 1985. New more attractive packaging was developed that extended shelf life of RTS orange juice, and product line extensions containing higher pulp and added calcium were introduced, etc. As the demand for convenience continued to grow, RTS orange juice became the preferred product form, overtaking FCOJ in market share by 1985/86.

Tropicana's RTS orange juice, Pure Premium, during the off-season was from single strength juice frozen into blocks and later thawed and pasteurized for packaging. It cost seven times as much to store and distribute as Recon, which was made by reconstituting bulk FCOJ, mostly in the major markets where most consumers lived. Tropicana had to charge more for its Pure Premium to cover its cost, but management believed Pure Premium had superior flavor to Recon because it had not been subjected to the heat of the evaporation process.

In order to distinguish Pure Premium from that of other brands and from a growing number of private and packer label RTS products that it was losing share to because of Pure Premium's higher price, Tropicana decided to tell the consumer that Pure Premium was not-from-concentrate and thus closer to fresh squeezed. Tropicana's marketing programs were highly successful in differentiating Tropicana Pure Premium as a unique orange juice that had never been concentrated or had water added, and consumers liked the NFC.

In spite of its being over $1.00 per gallon more expensive to consumers than Recon and almost $2.00 per gallon more expensive than FCOJ, between 1989 and 1998, NFC doubled its share of the U.S. orange juice market. Between 1985 and 1990, marketing successes with Pure Premium catapulted Tropicana to more than double its size. The Seagram Company, Ltd., a major marketer of distilled spirits and wines, spotted Tropicana's success with Pure Premium in the 1980's. Desiring to diversify into the juice business, which fit well with consumer trends toward health and fitness, Seagram purchased Tropicana from Beatrice for $1.2 billion in 1988. Seagram also felt that Tropicana's successes with Pure Premium could be expanded into untapped international markets, an arena where Seagram was an experienced player.

The three national orange juice brands continued their aggressive marketing activities and by 1990, brand advertising totaled $71 million, more than four times that of generic and a 417 percent increase over 1980 levels. These increased marketing efforts enabled the national brands to grow their share of the orange juice market from 33 percent in 1980 to 48 percent by 1990. Moreover, by 1990, these three national brands' share of the more value-added RTS orange juice segment had grown to 68 percent.

The supply disruptions from severe freezes in the 1980's also heightened the importance of less commodity-like juice products to the national brands. And high orange juice prices created by freeze reduced supplies opened a window of opportunity for juice beverages and juice blends to compete more directly with orange juice. These juice beverages contained citrus and sometimes other juices, but were not 100 percent juice, which reduced their dependency on fruit crops as a source of supply. Although juice blends were 100 percent juice, they contained several different types of fruit juices and were thus less dependent on supply conditions for a single type of fruit.

Tropicana developed and successfully launched its Twister juice beverage products, Citrus Hill began marketing Sunny Delight and later purchased Hawaiian Punch, and Minute Maid continued to successfully market its Hi-C and 5-Alive juice beverages. Higher orange and grapefruit juice prices also increased demand for Ocean Spray's juice beverage products, enhanced the opportunity for Dole to develop and successfully market its lines of chilled ready-to-serve juice blends (new breakfast juices, pure and light) and created an opportunity for a number of packer labels and regional brands to sell juice blends and juice beverages. The result was that between 1980 and 1994, juice beverages and juice blends grew their share of the U.S. fruit beverage market from 24 percent to over 40 percent, while orange juice declined from 47 percent to 34 percent of this market. The U.S. consumer, it seemed, was continuing to consume more fruit juice products in spite of supply restrictions that limited growth in orange juice availability.

Despite a severely cold freeze in late December of 1989, Florida's orange crop rebounded sharply. Since most of the trees were now further south, the 1989 freeze destroyed oranges but left most of the state's trees with little damage. By the early-to-mid 1990's, Florida's inventory of orange and grapefruit trees was the largest in history, and with Brazilian orange tree inventories also at record levels, world orange supplies grew dramatically while fruit prices tumbled to record lows. Juice marketers scrambled to move the large and growing supplies into the marketplace.

As a grower-owned cooperative with many fresh fruit grower members and (prior to the mid 1990's) regional brand, Citrus World, renamed Florida's Natural Growers, had looked for ways to market the supplies of red grapefruit from its members. In 1989, the cooperative introduced the first NFC red grapefruit juice. It was a better tasting product than the other Recon grapefruit juices on the market, whose taste suffered from the heat of evaporation. Consumers liked Florida's Natural NFC grapefruit juice so well that Tropicana followed with its own Ruby Red NFC grapefruit juice in 1993, which was also successful. The Florida Citrus Commission raised quality standards for grapefruit juice and processors with concentrate-based Recon products improved blending to enhance taste, color, and mouth feel. "We've gone from putting out what was a tart product and made it more palatable to the consumer. Better flavor. Higher quality," said one processor. Florida's

Natural also introduced their NFC orange juice in 1989, which was successful as well.

Although P&G supposedly had developed a new superior orange juice in the 1980's, it was reportedly too costly to produce and distribute commercially. This left P&G to compete with FCOJ against Minute Maid and with RTS against Tropicana, both well-established brands, particularly in their key segments. In 1992, P&G announced that they were selling their citrus processing facilities and retiring their orange juice brand, Citrus Hill, which had about a 7-8 percent share of the U.S. orange juice market. P&G cited losses in the orange juice business and that orange juice no longer fit their corporate objectives as reasons for its exit. Florida's Natural Growers had already been a strong regional brand for decades with Donald Duck and national with the relatively small Blue Bird canned citrus juices. But with its new Florida's Natural NFC orange juice it soon captured much of P&G's market share, and that along with successes from its new NFC red grapefruit juice product line enabled Florida's Natural to become the third largest national citrus juice brand.

Under Seagram ownership Tropicana continued its aggressive growth as a growing base of consumers chose Tropicana Pure Premium and its various line extensions in preference to other orange juices. In spite of a severe supply disrupting freeze in 1989, Tropicana sales of Pure Premium, and total consumption of NFC, which was mostly Tropicana products, grew every year. In the early 1990's Tropicana successfully expanded Pure Premium sales into Europe and Japan, and in spite of an even higher price due to shipping costs from Florida (Europe and Japan do not have significant supplies of domestic juice oranges), consumers responded strongly in favor of Tropicana's NFC. Spotting Tropicana's successes, PepsiCo purchased the company in 1998 for $3.3 billion.

Recognizing the continued growth of NFC, The Coca-Cola Company (Minute Maid) entered the NFC category in 2001 with

Simply Orange, which was followed by Simply Grapefruit and other Simply juices, all successful product lines. By 2008, NFC had grown to over half of the U.S. orange juice market. Consequently, the key focus of all three national citrus juice brands is now NFC, whose aggressive marketing activities have increased their share from less than half to two-thirds of the U.S. orange juice market.

The Fresh Citrus Sector

The fresh citrus sector is comprised of citrus packinghouses that supply retail supermarket chains and gift fruit shippers that sell to individual consumers. Fresh citrus packinghouses pack and market about 10 percent of Florida's total citrus production. Although only about 6 percent of Florida's juice oranges are sold as fresh fruit, the major portion of Florida's grapefruit, tangerines, and navel oranges are sold fresh rather than for processing. While oranges are the most popular citrus fruit consumed in the United States, Florida's leading fresh citrus variety is grapefruit, accounting for half of Florida's citrus produced for the fresh market.

In 2014, there were 40 certified fresh citrus packinghouses in Florida, collectively packing almost 25 million cartons (2013-14 season) of fresh citrus annually. By the 2020-21season, that number had declined to 14 fresh citrus packers packing 9.3 million cartons. They are either cooperatives or large growers. Packinghouses receive oranges and grapefruit in large pallet bins holding about 900 pounds of fruit. Fruit that is split or rotten is discarded; Fruit deemed unacceptable for the fresh market due to ripeness, size, shape, appearance of peel, etc., is culled as eliminations and sent to processing plants (about 25-30% of fruit for the domestic market and 40-50% for export markets). Typical preparation steps for fruit for the fresh market are (1) applying fungicides (2) air-drying the fruit, (3) waxing the fruit to prevent it from drying out, (4) separating it into grades and sizes, (5) packing the fruit into either

42-pound or 45-pound cardboard cartons, and (6) shipping the fruit to market.

Most of the packinghouses have a sales staff/function and sell fruit directly to the supermarket chains or to citrus market operators and brokers. Virtually all fresh sales are free on board (f.o.b.) prices from the packinghouse. About a third of Florida's fresh citrus is exported to foreign markets. Canada is the largest market for Florida fresh oranges and tangerines while Japan and Western Europe are leading foreign markets for Florida grapefruit.

There are 64 gift fruit shippers in Florida. Most are vertically integrated into grove ownership and fruit packing. Retail fresh citrus outlets sell fresh oranges, grapefruit, tangelos, and tangerines; fresh squeezed orange and grapefruit juice; citrus candies; citrus jellies and jams; citrus recipe books; and other Florida souvenir-type novelties. They also pack and ship fruit, usually as gifts, to customers nationwide.

Three important elements of success to a gift fruit business are to consistently offer a large variety of high-quality citrus fruit at its peak flavor in the retail outlet; to be located within an easy drive of targeted customers; and to financially be able to spend the required years to build a gift fruit shipping customer base. Targeted customers are those who like to shop for unique things, love fresh citrus and can afford it, and are quality-conscious.

The most successful retail citrus outlets have many customers and operate from late October into April. In addition to in-store sales, significant revenues also are generated by telephone, internet, and mail orders. The typical gift fruit customer orders several shipments of fruit per year for themselves and their families and friends.

Retail citrus outlets tend to be privately held family firms, many handed down over several generations. While large corporations generally are not a good fit as operators of this type of business—due to the complexities of sourcing a large, varied inventory and dealing with the logistics requirements of a shipping program direct to individual consumers for perishable commodities—they may contract these smaller businesses as affiliates with whom they exchange orders for marketing and delivery services. Most retail citrus gift fruit shippers, however, are growers who have decided to also grade, pack, and sell their fruit in their own retail market outlet. It is the unique nature of this specialty business that attracts growers.

Many consumers tasted their first and/or best Florida citrus from a gift fruit box or bag, or a retail gift fruit outlet. Indeed, the commercialization of Florida citrus began with fruit stands and holiday gift fruit.

Recent Threats and Challenges

Since 2000, on the supply side, the Florida orange crop was battered by hurricanes Charlie, Francis, Jeanne, Wilma, Irma, and Ian, an atypically high-priced real estate market that diverted citrus groves to real estate development, citrus canker that spread out of control, and the emergence of the most devastating citrus disease known—Huanglongbing (HLB), or citrus greening. Citrus canker is a bacterial disease that causes lesions on leaves, stems and fruit, making the fruit, which is safe to eat, too unsightly to be sold. It is spread primarily by wind-driven rain and human contact. If uncontrolled, it will eventually render trees unproductive. If controlled by management practices, it reduces fruit yields by 5 – 15 percent, and increases production costs. The federally mandated citrus canker eradication program legally required the destruction of all citrus trees in a 1900-foot radius of an infected tree. In 2006, the canker eradication program was discontinued because it was deemed unsuccessful after hurricanes in 2004 and 2005 spread canker throughout Florida citrus groves. Between 1995 when the eradication program began and 2006, 10 percent of Florida's citrus acreage (87,000 acres) was destroyed as part of the canker eradication effort, at a direct cost of $1.3 billion.

Citrus greening originated in China about 100 years ago. It is thought to be caused by the bacterium, *Candidatus* Liberibacter asiaticus. Greening has seriously affected citrus production in a number of countries in Asia, Africa, the Indian subcontinent and the Arabian Peninsula. Wherever the disease appeared, citrus production has been compromised with the loss of millions of trees. Greening was first discovered in Florida near Homestead in 2005. It is spread by a tiny insect called the Asian citrus psyllid when it feeds on the leaves of infected trees. It first reduces fruit yields, then kills the tree. Fruit from trees infected with greening is safe to eat, but it has an inferior flavor. Thus, it is culled and not processed into juice or packed as fresh fruit.

By February of 2008, less than three years after the first positive HLB test on a Florida citrus tree, the disease had spread to all 32 of Florida's commercial citrus-producing counties. Since 2009, the Florida citrus industry, federal and Florida state governments have invested approximately $170 million in 470 research projects to find a cure for HLB. But as of 2022, a cure has not been found.

Greening has reduced Florida's citrus crops from abundance to scarcity. Between the 2003/04 season, the year before greening was discovered in Florida, and last season (2021/22), orange production plummeted from 242 million boxes to 41 million boxes, while grapefruit production dropped even more, from 41 million boxes to 3 million boxes. In addition, Florida's citrus production base was substantially eroded. Between 2004 and 2021, the number of orange trees declined by 34 percent and the number of grapefruit trees by 75 percent.

On the market side, reduced production has brought about much higher prices for fruit and juice. Between 2003/04 and 2021/22, prices of oranges for processing more than tripled, and prices of grapefruit for processing more than quadrupled. During this same period, retail prices for orange juice went up by 82 percent while grapefruit juice prices increased by 104 percent. The result of these higher prices has been declining citrus juice consumption. Between 2003/04 and 2021/22, the retail orange juice market dropped by 49 percent and the grapefruit juice market by 126 percent.

Not all of these market declines are because of higher prices, however. Some nutritionists and physicians have been downplaying orange juice consumption because of its high sugar content, which they claim contributes to obesity. Orange juice does contain natural fruit sugars, the same as any 100 percent fruit juice. But non-diet soft drinks contain more sugar per ounce than orange juice, it is added sugar from other sources, and has no substantial nutritional benefits other than calories. Per capita consumption of non-diet soda is ten times as high as orange juice consumption, but health care professionals are not attacking soft drinks as much as orange juice. Also, orange juice has more nutrients per calorie than most other 100 percent fruit juices such as apple, grape, pineapple, and prune juice.[7]

Other nutritional attributes of orange juice that leading beverages such as soft drinks, other 100 percent fruit juices, coffee, tea, and milk do not have is that orange juice is unusually high in vitamin C and contains potassium citrate, vitamins or other compounds that may act as anti-oxidants in the body, flavonoids, folate, and thiamine. These attributes may contribute to improving blood vessel functioning, reduce the risk of some birth defects, reduce LDL (the bad cholesterol), help prevent kidney stone formation, and may have beneficial effects on blood pressure, which can help reduce the risk of cardiovascular disease. These attributes may also help support the immune system, and protect cells against free radicals, which could damage cells and tissue that could contribute to

---

[7] Rampersaud, Gail C. **"A Comparison of Nutrient Density Scores for 100percent Fruit Juices."** *Journal of Food Science*, Vol. 72, No. 4, 2007.

the development or progression of chronic disease. Importantly, some of these benefits, particularly those related to cardiovascular disease markers, have been demonstrated in clinical trials with orange juice.[8] [9] [10] Moreover, epidemiological research has shown that consuming orange juice compared to not consuming is not associated with an increase in bodyweight in children or adults.[11] [12] Unfortunately these messages have not been effectively conveyed to consumers.

Simple logic can dispel the myth that the high sugar content of orange juice contributes to obesity. During the 1970s and 1980s, good tasting diet sodas were introduced into the US marketplace. They became mainstream as consumers switched from caloric to

---

[8] Basile, LG, Lima, CG, Cesar TB. **"Daily Intake of Pasteurized Orange Juice Decreases Serum Cholesterol, Fasting Glucose, and Diastolic Blood Pressure in Adults."** *Proc Fla State Hort Soc.* 2010; 123: 228–233.

[9] Cesar, TB, Aptekmann NP, Araujo MP, Vinagre CC, Maranhao RC. **"Orange Juice Decreases Low-Density Lipoprotein Cholesterol in Hypercholesterolemic Subjects and Improves Lipid Transfer to High-Density Lipoprotein in Normal and Hypercholesterolemic Subjects."** *Nutrition Research.* 2010; 30 (10): 689-694.

[10] Morand C, Dubray C, Milenkovic D, Lioger D, Martin JF, Scalbert A, Mazur A. **"Hesperidin Contributes to the Vascular Protective Effects of Orange Juice: A Randomized Crossover Study in Healthy Volunteers."** *Am J Clin Nutr.* 2011; 93 (1): 73-80.

[11] O'Neil CE, Nicklas TA, Rampersaud GC, Fulgoni VL 3rd. **"One Hundred Percent Orange Juice Consumption is Associated With Better Diet Quality, Improved Nutrient Adequacy, and no Increased Risk for Overweight/Obesity in Children."** *Nutrition Research*; 2011; 21(9): 673-682.

[12] O'Neil CE, Nicklas TA, Rampersaud GC, Fulgoni III VL. **"100percent Orange Juice Consumption is Associated With Better Diet Quality, Improved Nutrient Adequacy, Decreased Risk for Obesity, and Improved Biomarkers of Health in Adults."** National Health and Nutrition Examination Survey, 2003-2006. *Nutrition Journal;* 2012; 11:107 (December 2012).

non-caloric soft drinks. Currently, diet sodas account for almost half of soft drink consumption. If beverages containing sugar were the cause of obesity, almost half of the American population would have lost weight when they switched to diet sodas, but this didn't happen. According to the Centers for Disease Control, between 1990 and 2015, the percentage of obese adults increased from 11.1% to 29.3%. Why would switching from consuming orange juice to consuming water or lower calorie beverages prevent consumers from becoming obese when switching to diet sodas didn't?

When or if scientists discover a cure for HLB, can the declining U.S. orange and grapefruit juice markets be turned around into growing markets that will support high enough prices for Florida growers to take the risks and commit the capital to replant the Florida citrus industry's 400,000+ acres of infected groves? Can Florida's $6.8 billion citrus industry be saved, and along with it the 33,300 jobs it supports?

The Commercial Florida Vegetable Industry

History

For over a hundred years, commercial vegetable production has been important to the economy of Florida. Commercial production began in central Florida in the mid-1800s and in southern Florida in the late 1800s. Some of the crops being grown were corn, rice, sugarcane, sweet potatoes, Irish potatoes, and other vegetables of all kinds. Produce was sold locally due to the difficulty of getting highly perishable products to northern markets in good condition. Large-scale vegetable production in Florida was made possible by the improvement of transportation, the development of production and marketing technology, and the adoption of refrigeration and other techniques for quality maintenance. in the north. Improved handling, preparation, and refrigeration permitted highly perishable items to arrive in the northern cities at the peak of quality and condition.

Between the Civil War and World War I, transportation by railroads became more readily available. This allowed many growing areas to move their produce quickly to northern markets. A major reason for the growth of railroads in Florida was the Internal Improvement Act passed by the state legislature in 1855. This Act offered cheap or free public land to investors, especially for transportation. It encouraged people like Henry Flagler and Henry

Plant to build railroads in Florida. Fast transportation put Florida vegetable production areas only two to five days away from the major city markets. The Act also provided the initial monies to drain the southern part of the state to make farmland.

Water is another asset that helped determine the location of early vegetable production areas. In 1940, Sanford-Oviedo area growers credited their success in growing celery to the fact that their farms were tiled for irrigation and drainage and that they had artesian wells. Two big water projects were initiated to help control water and allow these areas to become productive farmland. The first project began in 1912, resulting in the creation of the Everglades farming area by draining the Everglades. By 1951, the land surface had sunk as much as 7 feet. This led to a plan to save the area by controlling water levels. The other project was the "1949 Central and Southern Florida Flood Control District Project" to control both waters going into Lake Okeechobee from the north and discharge from the lake to the south.

Much of the land south of Lake Okeechobee is muck soil. Technically, muck is not soil, it is decayed vegetative matter from decades of being a lake bottom. Organic matter, commonly peat, forms the basis of muck soil. Muck has a high concentration of tannins which gives it a distinct dark black color. Plant yields on muck soils are higher than on most other soils. If not covered with water, the muck dissipates at the rate of 1.1 inch per year. However, most south Florida vegetable farming operations also include sugarcane, which calls for flooding the fields every so often to control insect pests.

To the northeast of Okeelanta, in what is now the downtown business area of Pahokee, farmers were protected by a high natural ridge. In February 1917, farmer J. R. Poland was ready to harvest his peas, beans, peppers, and potatoes. He heard that a frost had damaged crops throughout Florida. But Poland's crops were not affected because the high ridge protected the farm. Other farmers in the area were also unaffected and made large sums of money for their cabbage, beans, and tomatoes. When word spread that the vegetable fields had not been damaged in Pahokee, people started buying muck land to establish farms in the area. By 1920, there were 624 farms on the eastern side of Lake Okeechobee with over 12,000 acres of crops.

In the 1930s, the Florida Experiment Stations found that the muck soils could be made more productive by adding micronutrients to the soil. Celery crack stem could be prevented by the addition of boron. New insecticides and fungicides released in the 1940s helped control many problems. This allowed growers to increase the acreage of crops such as sweet corn since corn earworm could finally be controlled.

Starting in 1910, Dania (now Dania Beach), just south of Fort Lauderdale, nicknamed the "Tomato Capital of the World," was shipping 1,000 to 1,500 train carloads during the 3-month season. The area went out of production due to salt water intrusion following a hurricane in 1947 and construction of the "Dania Cut-off Canal" that same year. In the 1920s, Homestead shipped up to 35 train carloads of tomatoes per day.

Beginning in the 1940s, about 20,000 acres of wetlands along the north shore of Lake Apopka were drained and diked to create farmland in the lake's rich marsh soil. This "muck" soil, in addition to a long growing season, produced a variety of vegetables, making the area one of the top vegetable producing regions in the United States.

The availability of labor has historically been a concern of Florida growers. During World War II, there was a shortage of labor. In 1943, the U.S. Department of Agriculture brought in workers from Mexico. In 1944, the Florida Vegetable Committee coordinated a program to use offshore labor, war prisoners, and migratory domestic labor to harvest crops. During World War II, women worked on the farms, helping run the operations and harvest the crops. In 1947, growers depended on contract labor from the Bahamas.

Spray equipment was modernized and covered a greater area at one time. In 1929-30, the Vowell sprayer was introduced. It was mule drawn with a 7.5-gal spray pump and could spray two rows at a time. Prior to that, growers had to use knapsack sprayers. By the end of World War II, sprayers were still pulled by mules, but were air-cooled, engine-powered 15- gallon pumps that could spray four rows at a time. In October of 1946, late blight Phytophthora infestans devastated the fall tomato crop, but Dithane®, which had been introduced in 1944, saved the crop.

Florida vegetable producers have faced foreign competition for over 60 years. South Florida vegetable growers faced competition from Cuba with the importation of similar products during the winter and spring months. Foreign competition for the U.S. tomato market came from Cuba and Mexico and is still an issue with Mexico.

In conclusion, whether it is weather, pests, labor, transportation or foreign competition, the problems growers face today are similar to the problems that have challenged Florida vegetable growers for the last 100 years. With improved technology, expanded markets, and ingenuity, they have met these challenges and the results have in turn shaped the industry we know today.

Commercial Vegetable Production in Florida Today

In 2021, there were 26 vegetables produced in 23 states. Florida is one of the leading vegetable producers in the U.S. in terms of production and value. Production occurred on approximately 173,100 acres and generated $1.37 billion in gross sales in 2021. Florida's warm winter climate allows Florida to serve as the main vegetable supplier to the United States during late fall, winter, and early spring months. A national ranking of Florida's vegetable industry is not possible after 2016, because in subsequent years, some statistics were withheld to avoid disclosing data for individual operators. However, in 2016, Florida's vegetable sales ranked second in the U.S. (Dittmar, et.al.). In that year, Florida ranked first in the sales of tomatoes, snap beans, watermelons, and cucumbers.

Florida Vegetable Production and Value, 2021

Vegetable	Planted Acres	Production Tons	Value 1,000 $
Tomatoes	23,000	138.6	323,534
Strawberries*	9,900	35.0	239,780
Bell Peppers	11,200	63.7	152,669
Sweet Corn	34,400	75.0	208,165
Snap Beans	27,000	22.5	64,243
Watermelons	26,600	203.3	192,372
Squash	9,000	17.0	35,313
Cabbage	8,900	49.0	45,339
Cucumbers	21,500	91.2	94,879
Cantaloupe	1,600	7.5	9,563
Total	173,100	702.8	1,365,857

*These data are for 2020

USDA, National Agricultural Statistics Service, Southern Region News Release. *Vegetables*. Athens, Georgia February 17, 2022

Courtesy University of Florida IFAS

Courtesy University of Florida IFAS

The Florida Sugarcane Industry

History

Sugarcane was brought to the nation's oldest city, St. Augustine, by Spanish colonial governor Menendez in 1565. But commercial production of sugarcane did not begin until 1767 in Florida in the New Smyrna colony, and it was disrupted by the American Revolution. Attempts following the end of the revolutionary war to develop commercial operations failed in the 1800s because of unsuitable soils and inclement weather. This failure to launch sugarcane in central Florida moved pioneers further south during the late 1800s and early 1900s. Finally, in the 1920s the sugarcane industry was established around the southeastern shore of Lake Okeechobee.

Two events account for the growth experienced in the following decades: (1) the development of varieties suitable for the semi-tropical climate of the area and (2) the establishment of a water management system built to remove flood waters from agricultural lands during the wet season. Prior to 1960, there were three raw sugar mills, producing up to 175,000 tons of raw sugar annually from about 50,000 acres. The opportunity for further expansion came in 1960, when the United States stopped the importation of Cuban sugar and acreage restrictions on domestic cane were lifted by the federal government. In the following four years, acreage increased to 223,000, sugar mills to eleven, and sugar production to 572,000 tons. The acreage restrictions imposed again in 1965 were responsible for the closing of three mills, halting the industry's growth. With the expiration of the Sugar Act at the end of 1974 and the elimination of production controls, the Florida sugar industry continued its expansion trend.

Charles Stewart Mott was responsible for establishing U.S. Sugar, one of the country's most successful agribusiness operations. When he exchanged his ownership of the family bicycle and automobile tire business for General Motors stock, he became a major stockholder of GM. That earned him a slot as GM Vice-President and a position on its Board of Directors. In 1931, he purchased the land and other assets of a failed sugarcane farming and processing company located in Clewiston, Florida, on the southern tip of lake Okeechobee. He renamed the company United States Sugar Corporation and became its chairman. Mott convinced others that growing sugarcane in the rich muck soils around Lake Okeechobee was profitable. Florida sugar eventually replaced the sugar once imported from Cuba, leading to the sugarcane industry we know today.

The Industry Today

Today there are four raw sugar mills. One of these mills is owned by a corporation, two are owned by a family-owned company and one is cooperatively owned by growers. U.S. Sugar's Clewiston mill can process up to 42,000 tons of sugarcane per day. Recently there has been a push to expand sugar mills to produce refined sugar, which has a higher value than raw sugar. There are currently two refineries in the Glades, one that is totally integrated into U.S. Sugar's Clewiston Mill and one that is attached to Florida Crystal's Okeelanta Mill.

Sugar Production

Currently, Florida leads the nation in the production of sugarcane, producing 52 percent of the value of sugarcane for sugar. Louisiana and Texas are the other states producing sugarcane. Approximately 70 percent of Florida's commercial sugarcane production happens in South Florida near the southern tip of Lake Okeechobee, which is in Palm Beach County. The remaining 30% of Florida's sugarcane is grown in Hendry, Highlands, Glades, and Martin counties. Sugarcane planting takes place from late August through January. Sugarcane is harvested between late October and mid-April. Most of the fibrous portion of the cane stalks (called bagasse) is burned as fuel for sugar mills. United States sugar production falls short of U.S. sugar consumption, so almost one-fifth of the sugar consumed in the U.S. is imported. Brazil, India, Thailand, China, and the U.S. are the world's five largest producers of sugarcane in that order.

Sugar is an essential ingredient in our nation's food supply that's made from two different crops – sugarcane and sugar beets. Sugar is unique among major foodstuffs in that the identical commodity comes from two completely different crop systems. Sugarcane is grown in the warm climate of the tropics while sugar beets are a temperate zone crop, produced primarily in the industrialized countries of the northern hemisphere. Roughly 75% of the world's sugar comes from sugarcane, while the other 25% comes from sugar beets. In 2020, Russia, the United States, Germany, France, and Turkey were the world's five largest sugar beet producers. Wyoming, Montana, North Dakota, Colorado, and Nebraska are the major sugar beet producers in the U.S. Raw sugar is produced from processing the sugarcane crop, and then is further processed into refined sugar. Sugar produced from sugar beets is processed directly into refined sugar.

Sugarcane is a perennial crop. Thus, it doesn't need to be replanted every year. When sugarcane is harvested, it's cut just above the root level so new sprouts will grow, ready to be harvested again in 10–12 months. The cane plants grow to be 10–20 feet high. Sugarcane yields are highest in the year the cane is planted, and they decline slightly each year afterward. Subsequent crops after the sugarcane is first harvested are called stubble crops. Although the replanting decision depends on prices and yields, sugarcane is typically replanted after two or three stubble crops.

The sugar with which we are familiar is centrifugal sugar. The juice from sugarcane and sugar beets is boiled down and crystallized, and the mixture of crystals and liquid is poured into a centrifuge where the liquid is spun out, leaving sugar. Sugar has many important uses other than sweetening your coffee or tea. It is used in bakery products, dairy products, breads and cereals, salad dressings, jams and jellies, canned fruits and vegetables, soft drinks, processed meats and sausage, and candy.

While sugar is made from two different crops, the challenges to raise a crop and produce the sugar are often the same, from increased input costs to the ever-present threat of heavily subsidized foreign-produced sugar displacing domestic production. American sugar producers supply approximately 70 percent of America's needs; however, the U.S. still ranks as the world's third largest importer of sugar.

Sugar prices behave in cycles. Typically, a period of tight supplies and the resulting high prices results in an overexpansion in world sugar production. That leads to low world market prices, a small, lightly traded market. Because prices are low, investment in new production facilities lags while the low prices result in increased consumption. World consumption begins to exceed world production and stocks are drawn down. A poor beet crop somewhere sends prices to record levels, and the cycle continues.

United States Sugar Legislation

For many years it has been the policy of the U.S. government, for defense and strategic reasons, to preserve within the United States the ability to produce a substantial portion of our sugar requirements. The reason: Sugar is an essential and vital food product needed by American consumers, the supply of which on a worldwide basis has been marked by periods of alternating scarcity and surplus. It is unlikely that any significant quantity of sugar would be grown in the U.S. if American producers had to compete on the open world market with sugar produced using cheap tropical labor, or under subsidy in other countries.

In 1789, the new United States, seeking a means of supporting its government, imposed the first tariff on raw sugar to help raise revenue. This tariff also provided market protection to sugarcane growers in Louisiana after it became a U.S, territory in 1803. For years, production was afforded to our sugar producers solely through the tariff. Although the tariff did assist domestic producers, it still left them exposed to the volatile price fluctuations of the world sugar market. It also periodically increased the price of sugar to consumers in the United States without assuring them of adequate foreign sources of supply.

The first attempt to develop a new sugar program was made in 1933 by representatives of the sugar industry under authority of the Agricultural Adjustment Act. Finally in 1934, the Jones-Costigan Act went into effect. Its principal features were:

1. The determination each year of the quantity of sugar needed to supply the nation's requirements at prices reasonable to consumers and fair to producers.
2. The division of the U.S. sugar market among the domestic and foreign supplying areas by the use of quotas.
3. The allotment of these quotas among various processors in each domestic area.
4. The adjustment of production in each domestic area to the established quotas.
5. The equitable division of sugar returns among beet and cane processors, growers and farm workers

6 The levying of a tax on the processing of sugarcane and sugar beets, the proceeds to be used to make payments to producers to compensate them for adjusting their production.

In 1936, Congress ruled that the tax on processors described in feature number 6 was unconstitutional when used as a device to control production. So feature number 6 was omitted. In the Sugar Act of 1948, major changes were in the methods of establishing and allotting quotas. The Sugar Act was amended in 1951, 1956, 1961, 1962, 1965, and 1971. These amendments basically involved changes in sugar quotas and different price objectives. Section 902 of the Food and Agriculture Act of 1977 amended section 201 of the Agricultural Act of 1949 to provide that the price of the 1977 crop of sugar beets and sugarcane be supported through loans made to sugar processors or purchases of the processed products. If market prices of sugar did not equal loan prices plus accumulated interest, processors could turn over sugar under loan to the federal government without paying interest, and thus collect at least the loan price on the sugar. However, if the sugar was sold on the market, the loans had to be paid off plus accumulated interest.

Currently, sugar policy is contained in legislation known as the Farm Bill, which comes up for reauthorization every five years. The bill authorizes the USDA to offer loans on sugar being stored for consumers by America's sugar producers. Farmers use the loans on the sugar to pay their bills while sugar is being stored for their customers. Producers are paid throughout the year as customers take delivery and pay for the sugar. Sugar producers have nine months to pay back the loan. Because loans are repaid with interest and there are no subsidy checks, the policy operates at $0 cost. To meet America's sugar needs, USDA monitors the amount of sugar to be produced domestically along with sugar supplied from the more than 40 foreign countries given access, through trade deals, to import sugar at, or close to, duty free into the U.S. market.

If demand exceeds domestic production plus expected sugar imports, market access for additional foreign sugar can be increased to meet customer needs. Sugar policy does not allow for unlimited amounts of duty-free foreign sugar to enter the U.S. market. This helps provide American sugar producers with a more level playing field given the heavy subsidization of sugar production around the world.

Courtesy The Sugar Association

Florida Sugarcane Mill Courtesy U.S. Sugar

Raw Cane Sugar in A Warehouse, Courtesy Florida Sugar Cane League

The Florida Wholescale Landscape Nursery Industry

Spanish court historian Antonio de Herrera y Tordesillas tells us that the explorer Ponce de León chose the name Florida for two reasons: It was discovered during the Easter holidays, which the Spanish call *Pascua Florida*, or Feast of Flowers; *and* the land was lush and full of blooms.

The abundant plant life that so impressed Ponce de León has played an important role in Florida's subsequent history. By the 1700s Florida had become a popular destination for pioneering botanists and naturalists. English botanist Mark Catesby was collecting plants in Florida as early as 1731, and John Bartram, father of American botany, visited in 1765.

Florida was still a wild frontier when pioneering horticulturist Dr. Henry Perrine began experimenting in the Keys with edible and medicinal tropical plants. He grew and studied Florida's first avocados and mangos before he died in 1840.

One of Florida's earliest nurseries was Royal Palm Nurseries in Manatee County, established by Pliny and Egbert Reasoner in 1883. The brothers popularized grafting, developed the first pink grapefruit, and introduced many varieties of ornamental palms, acacias, and crotons to the United States.

Today, nursery and greenhouse products are Florida's leading agricultural commodities and Florida's landscape nursery industry is the second largest in the nation, with California ranking first. Florida, land of flowers, continues to live up to its name (Putnam, Adam).

Starting a Nursery Business

This section draws heavily on information from Sellmer and Dana

Introduction

Success in the nursery industry requires expertise in all phases of plant production, from propagation to post-harvest handling. It also requires the ability to handle employee management, financial management, marketing, and sales. Finally, all business ventures—including the nursery business—require a large capital investment to cover the cost of land, equipment, buildings, supplies, labor, and stock, in this case the initial purchase of plants for propagation and growing to salable size.

For those who have nursery experience and a strong knowledge of plants and plant culture, purchasing an existing business or starting a new one are logical options. It is easier to obtain financing for the purchase of an existing nursery. Consult a banker or lending agent for details. Those with limited experience can gain valuable experience by working for a reputable grower. Many well-established nurseries are always looking for good help as full-time, part-time, and seasonal employees.

The chance of successfully arranging financing from banks, private money-lending institutions, private investors, or government programs will be greatly enhanced with an organized, comprehensive business plan. A nursery business plan will help determine the type, market, location, site, design, and size of the operation. In addition, it will establish specific details on equipment and employee requirements, operational costs, estimated pricing schemes, projected return on investment, and asset availability.

Type of Business

There are three basic nursery formats:

- Landscape nurseries, which grow plants for retail sales and for their in-house landscape service;
- Retail nurseries, which include small nurseries and garden centers that grow plants on limited acreage for strictly retail sales to the homeowner and to landscape contractors; and
- Wholesale nurseries, which include contract propagators, contract growers, growers, and distributors (wholesalers) of nursery stock for wholesale to other nurserymen, landscape contractors, and retail outlets. In deciding on the type of nursery business, the type and form (rooted cuttings, seedlings, liners,[13] container grown, field-grown ornamentals, or a combination) of nursery stock that will be grown is determined.

Defining the Market

Defining the market involves answering the question: Who is the customer? This question will be partially answered by the type of business (landscape, retail, or wholesale) that is chosen and location of the operation (suburban, urban, or rural). For wholesale nurseries, the type of customer will depend partly on the characteristics of the nursery location and site (e.g., soil texture, temperature) as they affect the products to be sold (rooted cuttings, liners, balled and burlapped, balled and potted, or container grown ornamentals). The cost of shipping goods will determine a nursery's ability to compete in distant markets and thus affect its customer base. Retail and landscape nurseries, on the other hand, rely more on their proximity to large metropolitan areas and the

[13] Young plants that need to grow for another year in a greenhouse before being sold.

strength of the landscape industry to define the customer. Each nursery owner must analyze and organize his or her own marketing channels, develop a sales program, prepare the product for distribution, extend credit, and make collections. There are many resources available through bookstores and libraries which provide tips on marketing and business management techniques. Five possible approaches to defining the market and customers are:

- Define the types of customers within the target market (average salary, average number per household) based on census and local Chamber of Commerce data;
- Determine customer preferences for products and services through questionnaires and by investigating the local competition;
- Use local and national industry association data and commercial publications to determine market trends;
- Anticipate future preferences on the types and number of plants to grow; and
- Promote plants which you or other colleagues consider outstanding.

Choosing a Business Location

Location of the nursery may be predetermined based on availability of property or for personal reasons. Alternatively, there may be an opportunity to choose a location that best suits the needs and goals of the business. In either case, choice of location will directly affect decisions on marketing, employee hiring, and future expansion. In selecting a location, proximity to other nurseries, to major cities, and to highways should be considered. Other factors to consider include availability of transportation facilities, distance to market, competition, availability of reliable labor (full-time, part-time, and seasonal), and availability of inexpensive land for future expansion.

Another consideration in choosing a location is the present and future regulatory status of the area (e.g., zoning, taxation, labor relations, environmental issues, and product inspections).

Selecting a Specific Nursery Site

Site selection and layout are very important in the early stages of the planning process. The choices will directly affect success in growing and selling plants. Ideal land for the production of ornamental plants may not be available, so be prepared to make intelligent compromises. Prioritize a list of requirements based on what site characteristics are the most difficult and costly to repair after the land is purchased. Do not purchase land blindly because that will lead to greater cost, reduced plant quality, increased product losses, and reduced profit. As with purchasing a house, certain questions should come to mind. The following are some major topics and specific questions to consider.

- What Is the history of the site?
- If crop land, what crop was grown?
- If non-crop land, what was the land used for previously?
- Has the land been contaminated by pesticide or other toxic chemicals?
- If so, what chemicals and what reclamation efforts were employed?
- Are there underground structures on the property (e.g., utility lines, storage tanks, irrigation lines, and drainage tile)?
- If so, where are they located and what are their sizes and directions?
- Does the land drain poorly or flood? What soil type and characteristics are on the property? Soil maps are available from the Soil Conservation Service. Contracting for soil testing and

a soil survey will help determine soil quality and characteristics, which will be useful in determining nursery layout and production areas.

General characteristics for a good nursery soil include a tillable, well-drained soil with high water-holding capacity. The land should have less than a 5-percent slope to insure good air flow and surface drainage. Field production requires soils with good fertility, high organic-matter content (3-5%), and without obstacles (e.g., large stones, hardpans, or shale in the root zone). Balled and burlapped plants should be planted in a sandy-clay to clay-loam soil which will hold together in a ball. Bare-root plants require light-textured soils that are sandy-sandy loam in texture, can be worked easily, and will readily fall away from the roots as the plants are being harvested. Container production does not necessarily require good soil on site. Container growing medium can be brought from remote locations. However, good soil on site as a resource in media construction may be useful. A firm surface which allows for the placing of containers and ensures surface drainage is a must.

Source and Quality of Water

Of all the needs of plant production, the availability of ample, high-quality water is most important. Potential sources of water include lakes and ponds, streams, wells, and municipal water supplies. Each type has its own characteristics and problems. Lakes, ponds, and streams are the most inexpensive sources of water. Things to consider are the level of contaminants and debris, storage capacity to fulfill needs during the most demanding time periods (June-August), reliability, and the environmental and legal limitations which may be placed on natural water sources. Wells and municipal water sources tend to be expensive, may have quantity restrictions, and tend to have high ph. It is very important to determine water sources and limitations prior to purchasing land and starting in the nursery business. Since water quality is so vital, testing of the water quality prior to determining the nursery site is critical. Total soluble salts are acceptable between 175-525 ppm, with sodium levels between 20-40% of the total salts. However, the ideal level is less than 175 ppm salts and less than 35 ppm sodium. Adequate water volume is essential. Overhead sprinklers require a maximum of 10-acre feet of water for each acre of plants. Trickle irrigation systems are much more efficient than sprinklers. During hot, dry periods, weekly requirements may reach 3 inches per acre if using sprinklers. Automated irrigation systems are usually more reliable and consume less water and labor than manual methods. The amounts and distribution of rainfall also help determine irrigation needs.

Nursery Design and Space Requirements

The type and number of plants to be produced and the necessary facilities to produce and sell nursery product(s) will direct the decision on space allocation for the numerous activities within a nursery, which include propagation, production, storage, soil handling and potting, office space, sales, shipping and receiving, customer and employee parking, and roadways. The ultimate goal of estimating space requirements and nursery design is to maximize efficiency while minimizing capital and labor inputs. With this in mind, the layout should minimize the duplication of areas (e.g., potting and storage) and minimize the distance between work areas (e.g., propagation and growing). In starting the design process, you must decide whether you will propagate your own materials or purchase nursery stock (liners) to grow for sale as finished field- or container-grown plants. Propagation is a time-, money-, and space-consuming venture. Many nurseries choose to start by purchasing liners to grow on to finished size to quickly recover part of the original investment and to establish a customer base. Once this has occurred (after approximately 5 years), a plan can be implemented to establish propagation facilities (cold frames and greenhouses) with less financial and time risk. In designing the nursery, space should be left for the addition of propagation facilities, for specialized growing areas for propagules, and for future expansion. In all cases, when the size of growing areas and the number of plants expected to be grown are estimated, space for liner production, growing plants on as finished products, sales lots and walkways and roadways should also be estimated. Also include provisions for basic equipment and supplies in the initial planning stages. This will include the design and installation of an irrigation system, incorporation of injectors for fertilizer applications, tractors, trailers, trucks, soil sterilization equipment, pesticide spraying equipment, soil mixing equipment, chemical and tool storage, and office supplies and equipment.

Labor Requirements

The number of employees needed in the nursery will depend on the size of the operation. Starting with fewer personnel gives more time for training and requires less financing. Mistakes will also be less likely and less costly. The nursery business is inherently a seasonal activity with extended labor needs during the fall and spring. By diversifying the types of products (e.g., container-grown plants) and services (e.g., landscaping) offered, sales can be increased and down-time of seasonal and part-time labor (e.g., migrant workers and students) reduced.

Remember, personnel planning includes complying with Social Security, insurance, wage and hour, and income tax requirements for each of the employees.

Capital Requirements

The initial investment beyond land, buildings, equipment, supplies, and labor will also include utilities, insurance, interest on loans, taxes, licenses, and inspections. Plan carefully when establishing projected budgets and financial needs to include these expenses. There are many types of nursery operations, and choice of operation and the market plan will greatly affect the initial investment. Two to four years of operation are usually required before significant returns can be expected. And maximum returns may not be realized for 6 years or longer. Therefore, access to sufficient funds or credit to operate the business during the establishment period is critically important.

Description of a Florida Wholesale Landscape Nursery

Florida Wholescale Landscape Nursery and Greenhouse Producers primarily produce and sell live plants and provide landscape design and plant installation services. Their primary customers include homeowners, builders, apartments and condos, commercial businesses, mass merchandise stores, retail garden centers, and landscape contractors. Plant types produced and sold include tropical foliage plants, potted flowering or bedding plants, shrubs, palms, plant liners, deciduous shade trees, vines or ground covers, fruit trees, evergreen trees, and turfgrass. Florida is the number one producer of indoor house plants and tropical foliage in the country. In 2020, this industry had a total economic output impact of $31.4 billion and supported more than 266,000 jobs statewide (Farm Credit of Central Florida and Florida Nursery Growers and Landscape Association).

The following information is from a survey by Khachatryan, et.al.

General Characteristics

Most of the firms (86%) were established between 1980–2019. Twenty percent of firms were established between 2015–2019, 10 percent between 2010–2014, 22 percent between 2000–2009, 17 percent between 1990–1999, and 17 percent that were established between 1980–1989. Only 0.2 percent of Florida's wholescale landscape nursery and greenhouse producers were established before 1900, and 5.3 percent were established between 1900 and 1969.

Water Sources and Irrigation Methods

Groundwater wells supplied 57 percent of the water, followed by natural surface water (16 percent), city (municipal) water supplies (12%), recaptured sources (8 percent), and 3percent from reclaimed water sources. Irrigation water application methods used include overhead sprinklers (54 percent), drip irrigation (29 percent), sub-irrigation (ebb/flood, 4 percent), hand watering (8 percent), and other methods (13 percent).

Sales and Product forms

Almost half of the responding firms (46.4 percent) had less than $250,000 annual sales, 9.5 percent had sales between $250,000 and $999,999, 8 percent have between $1 to $4.9 million in sales and 2.8 percent have sales between $5 and $9.9 million. The other third has annual sales in excess of $10 million. Most product from sales (76.6 percent) were containerized; cut trees, budwood, scions, and seeds were 16.8 percent; balled/burlapped were 4.2 percent; and the remaining 2.4 percent were field grow bag, bare root, balled and potted/process balled, and in-ground containers/pot-in-pot.

Markets and Marketing Channels

Most firms (47.4 percent) sell their products through the re-wholesale market channel. This channel is followed by landscape firms

(26.2 percent), home centers (9.6 percent), single location garden centers (8.5 percent), direct-to-consumer (4.5 percent), multiple location garden centers (2.8 percent), and mass merchandisers (1.1 percent). This is the first year the survey has included direct-to-consumer as a market channel option. Direct-to-consumer sales could continue to increase as technology becomes more widely used by consumers and the green industry begins to be visible on more channels.

Courtesy State Archives of Florida

The Author, Cultivating Nursery Plants at Southern States Nurseries in 1970

Courtesy Mid Florida Nurseries

Courtesy Florida Nursery Mart

Courtesy Harris Citrus Nursery

On This Vacation

Drive beyond the theme parks, past the souvenir shops, the all-you-can-eat buffets, and chain hotels and you'll find a different Florida. In a way, it's a lot like Florida was before all of the theme parks and tourist attractions. It's a world of row upon row of citrus groves, which when in bloom during the spring, fill the air with the sweetest perfume-like scent; a world of pristine native range, and lush green cattle pastures with contented cows grazing in the distance; and 8,400 miles of coastline (second only to Alaska) consisting of broad coastal wetlands, vast estuaries, coral reefs, bays and inlets meandering past barrier islands, and over 800 miles of some of the most beautiful beaches in the world. Tour a Florida citrus grove; a cattle ranch; rent a canoe and paddle down one of Florida's wilderness rivers; charter a boat and take a trip along Florida's scenic intracoastal waterway; or just sit at a dockside restaurant on Florida's gulf coast and watch the red sun burn its way into the ocean as sunset turns into night.

References

Introduction

University of Florida / IFAS. Economic Impact of Agriculture in Florida 2022.

Rockport Analytics. 2019 Contribution of Travel and Tourism to the Florida Economy.

The Construction Association. The Economic Impact of Construction in the United States and Florida (in 2020).

The Turpentine Industry

Outland, Robert B. *Tapping the Pines: The Naval Stores Industry in the American South*. Louisiana State University Press, Baton Rouge, 2004.

The Florida Timber Industry

Almanac. **History: Timbering in North Florida**. Florida Natural Areas Inventory, Tallahassee, FL. 2005.

Composite Panel Association

Florida Forestry Association, **About Florida Forestry.**

Florida Land Steward, Timber Management. **Harvesting Methods. Fire.** University of Florida, IFAS Extension, 2021.

Forestryequipmentguide.com **Overview of Forestry Equipment**. January 2, 2019.

Howell, Michael. **Florida's Timber Industry – An Assessment of Timber Product Output and Use, 1993.** USDA, Forest Service, Southern Research Station. Resource Bulletin SE-153. Asheville, NC May 1995.

New Hampshire Dept. of Resources and Economic Development, Division of Forests and Lands and UNH Cooperative Extension. **Good Forestry in the Granite State: Recommended Voluntary Forest Management Practices for New Hampshire. 3.1. Timber Harvesting Systems.** 2010.

Nix, Steve. **Primary Timber Products Harvested When Selling Trees.** Treehugger. March 29, 2017.

University of Florida, IFAS. **Florida's Agriculture and Food System. Fast Facts 2021**.

The Florida Cattle Industry

Florida Cattlemen's Association, *www.floridacattlemen.org*.

Florida Department of Agriculture and Consumer Services, Division of Animal Industry. *Florida's Cattle Industry*. Tallahassee, FL, March 2007.

Florida Memory Project. *Florida Cattle Ranching*. Division of Library & Information Services, Florida Department of State.

Lacy, R. Curt, Carole H. Knight, and John L. McKissick. *Profitable Cattle Marketing for the Cow-Calf Producer*. University of Georgia Extension, Agricultural and Applied Economics Department, Athens, Georgia, 2012.

Parish, Jane A., Justin D. Rhinehart, and John D. Anderson. *Marketing Feeder Calves*. Mississippi State University Extension Service, Mississippi State, Mississippi, 2009.

The Florida Commercial Fishing Industry

Adams, Chuck. *The Commercial Bottom Trawling Industry in Florida: Balancing Environmental Impact with Economic Contribution.* Electronic data Information Service (EDIS) FE 345, Food and Resource Economics Department, Institute of Food and Agricultural Sciences, University of Florida, August 2002.

Hodges, Alan, David Mulkey, Effie Philippakos, and Chuck Adams. *Economic Impact of Florida's Commercial Fisheries and Aquaculture Industries.* Electronic Data Information Service (EDIS) FE 280, Food and Resource Economics Department, University of Florida, Institute of Food and Agricultural Sciences, October, 2002.

Florida Department of Agriculture and Consumer Services. *Florida Seafood and Aquaculture Overview and Statistics – Seafood.*

Florida Fish and Wildlife Conservation Commission, Fish and Wildlife Research Institute. Gulf of Mexico Fishery Management Council. Implemented Fishery Management Plans & Amendments for Reef Fish, Shrimp, Stone Crabs and Spiny Lobsters. Tampa, Florida.

National Oceanic and Atmospheric Administration, Office of Sustainable Fisheries, *Regional Fishery Management Councils.* Silver Spring, Maryland.

The Florida Citrus Industry

A.C. Nielsen. *Nielsen Data for U.S. Grocery Stores Doing $2 Million and Greater Annual Sales, Wal-Mart Stores Excluding Sam's Clubs, Mass Merchandisers and Drug Stores Doing $1 Million and Greater Annual Sales.* A.C. Nielsen, New York, NY, various issues.

FASS, USDA, NASS. *Florida Citrus Statistics, 2020-2021*. Florida Agricultural Statistics Service, United States Department of Agriculture, National Agricultural Statics Service, Orlando, FL, March 2022.

FDOC. *Citrus Reference Book*. Florida Department of Citrus, Economic and Market Research Department, Gainesville, FL 2021.

Florida Canners Association. *Year Book Season of 1961-1962, Processed Concentrated Orange Juice and Frozen Grapefruit Annual Packs in the State of Florida, Seasons 1940-41 to 1961-62* Inclusive. Florida Canners Association, Winter Haven, FL, 1963.

Florida Citrus Mutual. *Summary of Citrus Statistics*. Florida Citrus Mutual, Lakeland, FL, various issues.

The Commercial Florida Vegetable Industry

Dittmar, Peter J., Nicolas S. Default and Shinsuke Agehara. Chapter 1. **Commercial Vegetable Production in Florida.** University of Florida, IFAS Extension, May 2022.

Rosenberger, Stanley Eugene. **A History of the Florida Vegetable Industry and State Farmers' Markets for Vegetables.** Ph.D. Dissertation, University of Florida, August 1962.

Whidden, Alicia. **A History of Commercial Vegetable Production in Central and Southern Florida.** Proc. Fla. State Hort. Soc. 117:1-3. 2004.

The Florida Sugarcane Industry

American Sugar Alliance. **U.S. Sugar Policy – How It Works**. 2022

Florida Department of Agriculture and Consumer Services – Division of Plant Industry. **VIVA Florida 500: History of Sugarcane.** Gainesville, FL February 7, 2013

Palm Beach County Cooperative Extension – Agriculture. **Sugar Cane, Rice, and Sod.**

USDA, Economic Research Service. **U.S. Sugar Production**. 2021.

U.S. Sugar, current statistics.

U.S. Sugar. **History.**

Walton, Justin. **The 5 Countries that Produce the Most Sugar**. Investopedia, 2022.

The Florida Wholesale Landscape Nursery Industry

Farm Credit of Central Florida and Florida Nursery Growers and Landscape Association. **Florida's Nursery & Landscape Industry Generates Big Economic Impact.** March 2022.

Khachatryan, Hayk, Melinda Knuth, Alan Hodges, and Charlie Hall. **Florida Nursery and Landscape Industry Characteristics Report.** University of Florida, Food and Resource Economics Department, December 13, 2021.

Putnam, Adam. This Moment in Florida History: **Florida's Nursery and Horticulture Industry.**

Sellmer, James C. and Michael N. Dana. **Starting in the Nursery Business.** Department of Horticulture, Purdue University Coopera-tive Extension Service, West Lafayette, Indiana. March 1994.

About the Author

Allen was born in Tampa, Florida and raised in MacClenny, a small north Florida town approximately 30 miles west of Jacksonville. He is an agricultural economist with over forty years of experience in agribusiness. Allen currently owns and operates Morris Agribusiness Services, a consulting company. From December 2016 until June 2017, he was a consultant transitioning his responsibilities as Vice-President of Sales and Marketing for Blue Lake Citrus Products (BLCP) to Sun Orchard, the juice company that acquired BLCP in November 2016. BLCP produced and marketed the Noble brand of high-end specialty citrus juices as well as bulk citrus juices for other brands and labels, sold in supermarkets and restaurants nation-wide and in Canada. From 2007 until 2012, he was on the faculty of the University of Florida in the Food and Resource Economics (Agricultural Economics) Department. His responsibilities included both educational programs and research. Prior to that, Allen was Vice-President of Real Estate for Prudential Agricultural Investments, the largest lender to Florida agribusiness, and one of the largest agricultural realtors in Florida; Director of Procurement, Sales, and Industry Relations for Cutrale Citrus Juices USA, the North American subsidiary of Sucocitrico Cutrale, one of the world's largest citrus growers, processors and exporters, based in Araraquara, Brazil; Director of Procurement and Operations Planning for Tropicana, one of the largest global citrus juice brands; Manager of Operations Planning for The Cola Company; and Corporate Economist for A. Duda and Sons, Inc., a Florida-based family-owned company that, when Allen worked there, produced and marketed vegetables, sugarcane, cattle, citrus, and turf grass, with operations in Florida, Texas, California, Arizona, Idaho, Missouri and Australia. His advice on agribusiness has been sought after as a consultant by global soft drink companies, national and regional citrus juice brands, international agricultural asset management firms, commodity trading exchanges, insurance companies, banks, and citrus growers, processors, and exporters. His clients have ranged from Fortune 100 companies to family-owned businesses located in Winter Haven, Lake Wales, Naples, Bradenton, and Orlando, Florida; Houston, Boston, Visalia, California, Monterrey, Mexico, Sao Paulo and Matao, Brazil, Valencia, Spain, and London, England. Allen has been an invited speaker on agribusiness topics throughout Florida, in New York City, Boston, Houston, New Orleans, Los Angeles, Bialowieza, Poland, Dusseldorf, Germany, Oosterhout, Holland, London, England, Valencia, Spain, Sao Paulo and Limeira, Brazil, Monterrey, Mexico, and Ganzhou, China. He has taught four case studies on agribusiness in executive education programs at the Harvard Business School and been an invited speaker at international food policy meetings at Harvard University's Kennedy School of Government. He has published 37 peer-reviewed articles and four books about agribusiness topics and been an expert witness in four cases requiring the expert testimony of an agricultural economist.

Made in the USA
Monee, IL
03 November 2026

67383957R00066